Mechanics of Cellular Bone Remodeling

Coupled Thermal, Electrical, and Mechanical Field Effects

Mechanics of Cellular Bone Remodeling

Coupled Thermal, Electrical, and Mechanical Field Effects

Qing-Hua Qin

CRC Press
Taylor & Francis Group
Boca Raton London New York

CRC Press is an imprint of the
Taylor & Francis Group, an **informa** business

CRC Press
Taylor & Francis Group
6000 Broken Sound Parkway NW, Suite 300
Boca Raton, FL 33487-2742

First issued in paperback 2017

ISBN-13: 978-1-4665-6416-9 (hbk)
ISBN-13: 978-1-138-03371-9 (pbk)

Library of Congress Cataloging-in-Publication Data

Qin, Qing-Hua.
    Mechanics of cellular bone remodeling : coupled thermal, electrical, and mechanical field effects / Qing-Hua Qin.
        p. ; cm.
    Includes bibliographical references and index.
    ISBN 978-1-4665-6416-9 (alk. paper) -- ISBN (invalid) 978-1-4665-6419-9 (hbk.)
    I. Title.
    [DNLM: 1. Bone Remodeling--physiology. 2. Bone Remodeling--radiation effects. 3. Bone and Bones--metabolism. 4. Bone and Bones--physiology. 5. Osteoblasts--physiology. WE 200]

612.7'51--dc23   2012027636

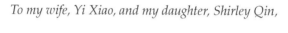

To my wife, Yi Xiao, and my daughter, Shirley Qin,

for filling my life with love and joy.

Contents

Preface

The bone remodeling process is an important phenomenon in biomaterials. It is a two-stage process carried out by teams of cells known as basic multi-cellular units, and it involves the removal of mineralized bone by osteoclasts (the cells with functions of bone resorption) followed by the formation of bone matrix through the osteoblasts (the cells with functions of bone formation) that subsequently become mineralized. The basic multicellular unit is indeed a wandering team of cells that dissolves an area of the bone surface and then fills it with new bone. Bone remodeling is, in fact, viewed as serving to adjust bone architecture to meet changing mechanical needs and it helps to repair microdamages in the bone matrix to prevent accumulation of old bone. It also plays an important role in maintaining equilibrium between bone formation and resorption.

Research to date on bone remodeling has resulted in a great deal of new information and has led to improvements in design and biomedical practices. Articles have been published in a wide range of journals attracting the attention of both researchers and practitioners with backgrounds in the mechanics of solids, applied physics, applied mathematics, mechanical engineering, and materials science. However, no extensive, detailed treatment of this subject has been available up to the present. It now appears timely to collect significant information and to present a unified treatment of these useful but scattered results, which should be made available to professional engineers, research scientists, workers, and students in applied mechanics and biomedical engineering (e.g., physicists), and materials scientists.

The objective of this book is to fill this gap so that readers can obtain a sound knowledge of bone remodeling and its mathematical representation. This volume details the development of each of the techniques and ideas, beginning with a description of the basic concept of bone remodeling from a mathematical point of view. From there we progress to the derivation and construction of multifield and cellular bone remodeling and show how they arise naturally in the response to external multifield loads.

This book is written for researchers, postgraduate students, and professional engineers in the areas of solid mechanics, physical science and engineering, mechanical engineering, materials science, and biomedical engineering. Little mathematical knowledge beyond the usual calculus is required, although differential and integral representation is used throughout the book. It contains a comprehensive treatment of bone remodeling using linear multifield theory and various special solution methods. Our hope in preparing this book is to present a stimulating guide and then to attract interested readers and researchers to a new field that continues to provide fascinating and technologically important challenges. The reader will benefit from the thorough coverage of general principles for each topic, followed by detailed mathematical derivations and worked examples, as well as tables and figures where appropriate.

Chapters 1 and 2 provide a brief description of bone materials and linear bone remodeling theory in order to establish basic notations and fundamental concepts for reference in later chapters. Chapter 3 deals with problems of multifield internal bone remodeling, beginning with a discussion of linear theory of thermoelectroelastic bone and ending with a brief description of the solution for both homogeneous and inhomogeneous hollow circular cylindrical bone, as well as on the extension to thermomagnetoelectroelastic bones. Chapter 4 is concerned with another type of multifield bone remodeling: surface bone remodeling.

Chapter 5 describes a hypothetical regulation mechanism on growth factors for bone modeling and remodeling under multifield loading. Chapter 6 presents a description of the RANK–RANKL–OPG pathway and formulation for analyzing the bone remodeling process induced by parathyroid hormone (PTH) at the cellular level. Chapters 7 and 8 are an extension of Chapter 6 to the cases of mechanical and pulsed electromagnetic stimulus. Chapter 9 describes some experimental approaches on bone tissues.

I am indebted to a number of individuals in academic circles and organizations who have contributed in different but important ways to the preparation of this book. In particular, I wish to extend appreciation to my postgraduate students for their assistance in preparing this book. Special thanks go to Li-Ming Leong of Taylor & Francis for her commitment to the publication of this book. Finally, I wish to acknowledge the individuals and organizations cited in the book for permission to use their materials.

I would be grateful if readers would be so kind as to send reports of any typographical and other errors, as well as their more general comments.

Qing-Hua Qin
Canberra, Australia
Qinghua.Qin@anu.edu.au

The Author

Qing-Hua Qin received his BE degree in mechanical engineering from Chang An University, China in 1982, and his MS and PhD degrees in applied mathematics from Huazhong University of Science and Technology (HUST), China in 1984 and 1990, respectively. He joined the HUST Department of Mechanics as an associate lecturer in 1984 and was promoted to lecturer of mechanics in 1987 during his PhD candidature period. After spending 10 years lecturing at HUST, he was awarded a DAAD/K. C. Wong research fellowship in 1994, which enabled him to work at the University of Stuttgart in Germany for 9 months. In 1995 he left HUST to take up a postdoctoral research fellowship at Tsinghua University, China where he worked until 1997. He was awarded a Queen Elizabeth II Fellowship in 1997 and a professorial fellowship in 2002 (both by the Australian Research Council) at the University of Sydney and stayed there until December 2003. He is currently working as a professor in the Research School of Engineering at the Australian National University, Canberra, Australia. He was appointed a guest professor at HUST in 2000 and was a recipient of the J. G. Russell Award from the Australian Academy of Science. He has published over 200 journal papers and 6 monographs.

1

Introduction to Bone Materials

1.1 Introduction

Bone is a highly organized and specialized support tissue that is characterized by its rigidity, porosity, and hardness. It is a connective tissue with both cellular and extracellular matrix composed of extracellular fibers and a dense crystalline material. Impregnated with inorganic salts, bone contains mainly salts of calcium such as calcium phosphate, calcium carbonate, etc. The inorganic portion occupies two-thirds of the bone and the organic salt component constitutes the rest. Inorganic salts are mainly responsible for the rigidity and hardness that allow bone to resist compression caused by the forces of weight and impact. The organic connective tissue of bone makes it resilient, thus affording resistance to tensile forces. Bones support and protect various organs of the body, produce red and white blood cells, and store minerals.

1.2 Types of Bones

Living bones are classified on different bases. Generally, five types of classification are used. They are based on a macroscopic approach, microscopic observation, geometric shape, patterns of development, and the regions in which bones are located. In this section, we present a brief review of the first three classifications. The remaining two types, which are not commonly used, are listed in Appendix A for readers' convenience.

1.2.1 Bone Types Based on the Macroscopic Approach

From the viewpoint of bone macrostructure there are two primary types of bone: compact bone and spongy bone.

- *Compact bone.* Compact bone is also known as cortical bone. It is the hard material that makes up the shaft of long bones and the outside surfaces of other bones. Compact bone consists of cylindrical units called osteons (see Figure 1.1). Each osteon contains concentric lamellae (layers) of hard, calcified matrix with osteocytes (bone cells) lodged in lacunae (spaces) between the lamellae. Smaller canals, or canaliculi, radiate outward from a central canal, which contains blood vessels and nerve fibers. Osteocytes within an osteon are connected to each other and to the central canal by fine cellular extensions. Through these cellular extensions, nutrients and wastes are exchanged between the osteocytes and the blood vessels. Perforating canals (Volkmann's canals) provide channels that allow the blood vessels running through the central canals to connect to the blood vessels in the periosteum that surrounds the bone. In general the Young's modulus of compact bone is about 18 GPa and its porosity ranges from 5% to 30%.

- *Spongy bone.* Spongy bone, also called cancellous or trabecular bone, is an interior meshwork of trabeculae. Trabeculae consist of thin, irregularly shaped plates arranged in a latticework network (see Figure 1.1). They are similar to osteons in that both have

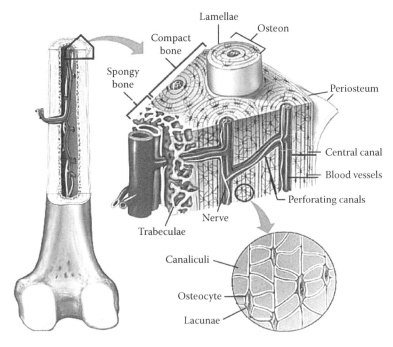

FIGURE 1.1
Illustration of compact and spongy bone.

osteocytes in lacunae lying between calcified lamellae. As in osteons, canaliculi present in trabeculae provide connections between osteocytes. However, because each trabecula is only a few cell layers thick, each osteocyte can exchange nutrients with nearby blood vessels. Thus, no central canal is necessary. The Young's modulus of spongy bone is only about 1 GPa, and the corresponding porosity is between 30% and 90%.

1.2.2 Bone Types Based on Microscopic Observation

On the basis of microscopic observation of matrix arrangement, bones can be divided as follows:

- *Lamellar bone.* The lamellar bone is a normal type of adult mammalian bone composed of thin plates (lamellae). Whether spongy or compact, it is composed of parallel lamellae of bony tissues in the former and concentric lamellae around a vascular canal in the latter. Lamellar organization reflects a repeating pattern of collagen fibroarchitecture (Figure 1.2a).

- *Woven bone.* Woven bone is immature bone containing collagen fibers arranged in irregular random arrays. It has smaller amounts of mineral substance and a higher proportion of osteocytes than lamellar bone (see Figure 1.2b). Woven bone is temporary and is eventually converted to lamellar bone; this type of bone is also pathological tissue in adults, except in a few places such as areas near the sutures of the flat bones of the skull, tooth sockets, and the insertion sites of some tendons.

1.2.3 Bone Types Based on Geometric Shape

On the basis of geometric shape, bones are classified into six classes: long bones, short bones, flat bones, irregular bones, pneumatic bones, and sesamoid bones.

- *Long bones.* Long bones include some of the longest bones in the human body, such as the femur, humerus, and tibia, but they are also some of the smallest, including the metacarpals, metatarsals, and phalanges. These bones typically have an elongated shaft and two expanded ends, one on either side of the shaft (see Figure 1.3). The shaft is known as the diaphysis and the ends are called epiphyses. Normally, the epiphyses are smooth and articular. The shaft has a central medullary cavity where the bone marrow lies. Long bones that have a longitudinal emphasis are used as levers. The upper extremity bones are lighter, as they are used for a greater

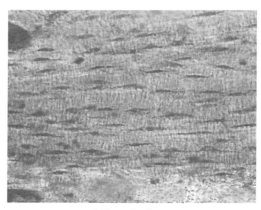

(a) Lamellar bone

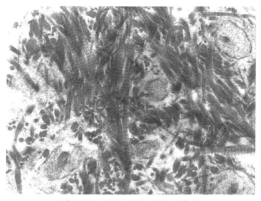

(b) Woven bone (non-lamellar)

FIGURE 1.2
Configuration of lamellar and woven bones.

range of motion—for actions such as manipulation of objects, and for speed (a boxer throwing a knockout punch). The femur and tibia are much thicker as they bear the body's weight. The fibula does not bear the weight of the body as it does not form a joint with the femur, and it is therefore much smaller. Because of its muscle attachments, it is extremely bumpy, especially in the competitive athlete. An additional type of long bone is the modified long bone. These bones have a modified shaft or modified ends. They have no medullary cavity such as that present in typical long bones. Examples of this class of bone are the clavicles and the bodies of vertebrae.

- *Short bones.* Short bones have comparably larger width than length (Figure 1.3) and have a primary function of providing support

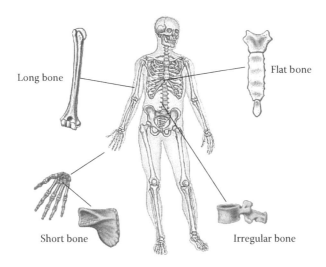

FIGURE 1.3
Illustration of bones classified based on geometric shapes.

and stability with little movement. They are roughly cube shaped and have only a thin layer of compact bone surrounding a spongy interior. They are found primarily in the feet and hands (carpal and tarsal bones). Further examples of this class of bone include the cuboid, cuneiform, scaphoid, and trapezoid bones, etc. Short bones are used for elasticity and adaptability. When walking, the feet must bear extreme amounts of weight and conform to endless varieties of terrain. The arrangement of the bones and tissue allows the feet almost to mold to these surfaces and to absorb shock.

- *Flat bones.* Flat bones are flat in appearance and have two prominent surfaces known as the anterior and posterior surfaces, which are formed from compact bone to provide strength for protection (see Figure 1.3). Their centers consist of cancellous (spongy) bones and varying amounts of bone marrow. These bones resemble shallow plates and form boundaries of certain body cavities. An example of a flat bone is the scapula (the shoulder blade), which protects several structures in the back region. In general, the main function of flat bones is to provide protection to the body's vital organs and to provide a base for muscular attachment.

- *Irregular bones.* As the name implies, the shape of these bones is irregular and they do not fit into any other shape category (Figure 1.3). Typical examples of irregular bones are the vertebrae, sacrum, and bones in the base of the skull. They consist primarily of cancellous bone, with a thin outer layer of compact bone. The vertebrae have a shape that allows them to protect the central nervous system, provide canals (foramina) for nerves to exit, and have flat bodies and other processes for joint formation, designed for extreme forms of movement yet bearing substantial amounts of weight. (For example, the lumbar region houses interlocking articular processes that allow an athlete to form the vital strong arch.) Irregular bones also have processes such as the spinous processes (the backbone, which one can see posteriorly) and transverse processes that allow muscles to attach and permit localized rotations.

- *Pneumatic bones.* Pneumatic bones, also called hollow bones, are hollow or contain many air cells—for example, the mastoid part of the temporal bone. Generally, pneumatic bones can also be categorized as irregular bones because of their irregularity in shape. There is a difference between the two types, however, that is very important, and for that reason they are often classified separately. The characteristic difference is the presence of large air spaces in these bones, which makes them lightweight. Pneumatic bones form the major portion of the skull in the form of the sphenoid, ethmoid (Figure 1.4), and maxilla. In addition to making the skull light in weight, they also contribute to the resonance of sound and function as air-conditioning chambers for inspired air.

- *Sesamoid bones.* Sesamoid bones are usually short or irregular, embedded in tendon and joint capsules. They have no periosteum and ossify after birth. They are related to an articular or nonarticular bony surface, and the surfaces of contact are covered with hyaline cartilage and lubricated by a bursa or synovial membrane. The most obvious example of sesamoid bone is the patella (kneecap) which sits within the quadriceps tendon (Figure 1.5). Other sesamoid bones

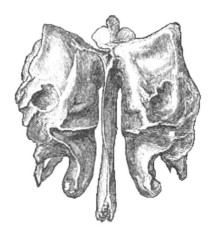

FIGURE 1.4
Illustration of ethmoid bones (http://www.learnbones.com/skull-cranial-and-facial-bones-anatomy; accessed March 2011).

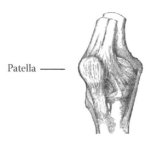

Patella ——

FIGURE 1.5
Illustration of patellar bones.

are the pisiform (smallest of the carpals) and the two small bones at the base of the first metatarsal. Sesamoid bones are usually present in a tendon where it passes over a joint, which serves to protect the tendon.

Bones can also be classified based on their pattern of development and their region. Details can be found in Appendix A.

As well, from a material standpoint, bone can be viewed as a composite material composed of about 65% hydroxyapatite crystals and about 34 type I collagen fibrils (by dry weight). Water in bone usually makes up 25% of the wet weight of bone material.

1.3 Bone Functions

To allow optimal function, the arrangement of individual bones is as precise, orderly, and purposeful as the full skeletal system itself, and their distribution from top to bottom is extremely balanced. Most of the bones in our body are structured in a symmetrical fashion. That is, many of our bones are matched on each side of the body. This matched design allows us to balance and stabilize ourselves in the face of various forces that act on our bodies. With optimal arrangement, the bone forms the basic unit of the human skeletal system, provides the framework for bearing the weight of the body, protects the vital organs, supports mechanical movement, hosts hematopoietic cells, and maintains iron homeostasis. These functions can be classified into three categories:

- Mechanical functions of bones
  - Protection. In numerous places inside the body, bones serve to protect important and delicate organs, reducing risk of injury to them. The best examples are the brain (protected by the skull),

heart and lungs (protected by the ribcage), and spinal cord (protected by vertebrae).

- Shape. Because of their rigid nature, bones provide a framework around which the body is built and is kept supported. Thus, bones are responsible for the shape and form of a human body.

- Movement. Working with skeletal muscles, tendons, ligaments, and joints, the bones form the moving machinery of the human body and function together to generate and transfer forces so that individual body parts or the whole body can be manipulated in three-dimensional space. One of the major roles of bones in movement is that they act as levers, which make use of the forces generated by skeletal muscles in a beneficial way (i.e., the bones of the upper and lower limbs pull and push, with the help of muscles). Leg and arm movements are the most obvious examples; a less obvious example is that ventilation of the lungs depends on movement of the ribs by skeletal muscles. Interactions between bones and muscles are studied in biomechanics.

- Sound transduction. Bones are important in the mechanical aspect of hearing. In fact, our ability to hear is largely dependent on bones. The arrangement of bones inside the ear structure is essential for audition.

- Synthetic functions of bones
  - Synthesis of blood cells. The major synthetic role of bone is in the production of blood cells. The bones themselves are not capable of doing this. Instead, they house the bone marrow, which contains hematopoietic stem cells capable of producing blood cells. In infants, the bone marrow of all long bones is capable of this synthesis; as a person ages, however, the red marrow turns into yellow fatty marrow, which is no longer capable of hematopoiesis. The red marrow in adults and older individuals is restricted to the vertebrae and the heads of the tibia and femur.

- Metabolic functions of bones
  - Mineral storage. Bones serve as an important storehouse of minerals such as calcium and phosphorus. In general, 97% of the body's calcium is stored in bone. Here it is easily available and turns over quickly. When required, bones release minerals into the blood, facilitating the balance of minerals in the body.
  - Fat storage. The yellow bone marrow of long bones stores fatty acids.
  - Marrow holder. This is secondary to the production of maximum strength for minimum weight: The cavities produced in unstressed areas (like the holes in the tubes of a bicycle frame) are used for marrow or, in some places (mastoids), just for air storage.

- Role in acid–base balance. Bones buffer the blood against excessive pH changes by absorbing or releasing alkaline salts.

- Storage of chemical energy. With increasing age some bone marrow changes from red to yellow. Yellow bone marrow consists mainly of adipose cells, with a few blood cells. It is an important chemical energy reserve.

- Growth factor storage. A mineralized bone matrix stores important growth factors such as insulin-like growth factors, transforming growth factor, bone morphogenetic proteins, and others.

- Detoxification. Sometimes, heavy metal and some foreign elements need to be removed from the blood. These are stored in the bones after such removal. Thus, their effects on body tissues are reduced. These stored wastes are removed naturally by excretion.

- Endocrine organ. Bones control phosphate metabolism by releasing fibroblast growth factor C 23 (FGF-23), which acts on the kidneys to reduce phosphate reabsorption. Bone cells also release the hormone osteocalcin, which contributes to the regulation of blood sugar (glucose) and fat deposition. Osteocalcin increases both insulin secretion and sensitivity, in addition to boosting the number of insulin-producing cells and reducing stores of fat.

1.4 Bone Cells

Bone cells in the body have specific functions, all of which are necessary for bones to develop and grow. They are also needed to retain the strength of bones. According to Wikipedia, there are five basic types of bone cell present in bone tissue, with functions including reaction of the body, fracture and trauma, and secretion of the chemical compound that bones are made of. Details of the five basic types of bone cells are as follows:

- Osteoblasts. Osteoblast cells, commonly called bone-forming cells, are derived from mesenchymal stem cells and are responsible for bone matrix synthesis and its subsequent mineralization. Mesenchymal stem cells are multipotent stem cells that can differentiate into a variety of cell types including osteoblasts, chondrocytes, and adipocytes. Zinc, copper, and sodium are some of the minerals required in the process of mineralization. Osteoblasts also secrete the basic compound needed to help in the process of bone repair, bone growth, and, in some cases, bone regrowth. In addition, osteoblasts produce bone tissue called osteoid from collagens and calcium, built outward from the marrow. Bone is a dynamic tissue that is constantly being reshaped by osteoblasts, in charge of producing matrix, mineral, and osteoclasts, which remodel the tissue. The number of osteoblast cells tends to decrease with age, affecting the balance of formation and resorption in bone tissue.

- Osteocytes. Osteocytes are mature osteoblasts, but they have a different function altogether. Some osteoblasts turn into osteocytes during the process of bone formation. Osteocytes situated deep in the bone matrix maintain contact with newly incorporated osteocytes in osteoid, and with osteoblasts and bone lining cells on bone surfaces, through an extensive network of cell processes (canaliculi). They are thought to be ideally situated to respond to changes in physical forces upon bones and to transmit messages to the osteoblastic cells on the bone surface, directing them to initiate resorption or formation responses. They also maintain metabolism and participate in nutrient and waste exchanges through the blood. In general, an osteocyte cell contains a nucleus and a thin ring of cytoplasm. When osteoblasts become trapped in the matrix they secrete, they become osteocytes. The space that an osteocyte occupies is called a lacuna. Although osteocytes have reduced synthetic activity and, like osteoblasts, are not capable of mitotic division, they are actively involved in the routine turnover of bony matrix through various mechanosensory mechanisms. They destroy bone through a rapid, transient (relative to osteoclasts) mechanism called osteocytic osteolysis. Osteoblasts/osteocytes develop in mesenchyme. Hydroxyapatites, calcium carbonates, and calcium phosphates are deposited around the osteocytes.

- Osteoclasts. Osteoclasts were discovered by Kolliker in 1873 [1]. An osteoclast is a large cell, 40 µm in diameter. Osteocytes contain 15–20 closely packed oval-shaped nuclei. They are found in pits in a bone surface, called resorption bays, or Howship's lacunae (a groove or cavity, usually containing osteoclasts, that occurs in bone undergoing reabsorption). They are large multinucleated cells, like macrophages, derived from the hematopoietic lineages. Osteoclasts function in the resorption of mineralized tissue and are found attached to the bone surface at sites of active bone resorption. Their characteristic feature is a ruffled edge, where active resorption takes place with the secretion of bone-resorbing enzymes, which digest bone matrix. They are the cells responsible for the breakdown of bone tissue. In particular, they release calcium and are much larger than other bone cells. Osteoclasts release an enzyme that destroys bone tissue when deemed necessary. These cells are an important part of bone growth and bone repair, and they also help in rebuilding bone when it is damaged. Osteoclasts and osteoblasts are together instrumental in controlling the amount of bone tissue: Osteoblasts form bones and osteoclasts resorb bones. Osteoclasts are formed by the fusion of cells of the monocyte-macrophage cell line (monocytes are a type of white blood cell). Osteoclasts are regulated by several hormones, including PTH from the parathyroid gland, calcitonin from the thyroid gland, and growth factor interleukin 6 (IL-6). The last hormone, IL-6, is one of the factors in the disease osteoporosis, which is an imbalance between bone resorption and bone formation. Osteoclast activity is also mediated by the interaction of two molecules produced by osteoblasts: osteoprotegerin and the RANK ligand. Note that these molecules also regulate differentiation of the osteoclast [1].

- Osteogenics. Osteogenic cells are responsible for trauma response. They are found in the bone tissue, which contacts endosteum and periosteum. These cells are also responsible for the healing process starting when trauma is experienced. Osteogenic cells heal by calling on other cells, such as osteoblasts (bone-forming cells) and osteoclasts (bone-destroying cells). When they all work together, they repair damage to the bones. It is noted that osteoblasts arise from osteoprogenitor cells located in the deeper layer of periosteum and the bone marrow. Osteoprogenitors are immature progenitor cells that express the master regulatory transcription factor Cbfa1/Runx2. Osteoprogenitors are induced to differentiate under the influence of growth factors, in particular the bone morphogenetic proteins (BMPs) [2]. Aside from BMPs, other growth factors including the fibroblast growth factor (FGF) [2], the platelet-derived growth factor (PDGF), and the transforming growth factor beta (TGF-β) may promote the division of osteoprogenitors and potentially increase osteogenesis.

- Bone lining. Lining cells are a type of bone cell that creates the lining of bones. These cells come from osteoblasts. Cell lining can be found in the surface of adult bones. Lining cells regulate calcium and phosphate in bones. These minerals must be regulated to maintain a healthy mineral content.

It should be mentioned that bone surfaces that are not under remodeling or modeling are covered by elongated, thin cells. These bone lining cells are thought to be either inactive osteoblasts, which perhaps can be activated to produce bone matrix, or a cell type of its own. In Gardner staining, osteocytes are seen as a row of nuclei along bone surfaces, since their cytoplasm does not contain many cell organelles and is therefore only weakly stained in this procedure.

An important aspect concerning bone lining cells is that the retraction or removal of these cells is a necessary step in starting osteoclastic bone resorption [3,4]. Retraction exposes the underlining osteoid to proteolytic enzymes of osteoblasts [5] and, further, to osteoclastic resorption. The fully mineralized bone matrix is exposed only after retraction of bone lining cells and removal of organic osteoid. Therefore, it can be assumed that bone lining cells are remarkable in vivo regulators of osteoclastic bone resorption. The exact mechanisms of this cascade are not yet known, but it seems that soluble factors as well as cell–cell contacts have a role in the activation of osteoclasts [6,7].

1.5 Osteoporosis

According to Osteoporosis Australia (http://www.osteoporosis.org.au), osteoporosis is a condition in which the bones become fragile and brittle, characterized by a decrease in the density of bone and leading to a higher risk of fractures than in normal bones. Osteoporosis leads to an abnormally porous bone that is compressible, like a sponge. It occurs when bones lose minerals, such as calcium, more quickly than the body can replace them, leading to a loss of bone mass or density. As a result, bones become thinner and less dense (see Figure 1.6), so even a minor bump or accident can cause serious fractures.

Osteoporosis is in fact a bone disease where bone mass decreases over time. Its manifestations are subdivided into postmenopausal and senile forms [8,9]. The disease process results from a net increase of bone resorption over deposition.

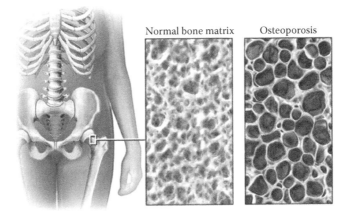

FIGURE 1.6
Comparison between normal bone matrix and osteoporosis in the human hip. (Picture from http://www.halanmediacorp.com/wp-content/uploads/2009/11/osteoporosis.jpg; accessed March 2011.)

It is reported that adult bone mass peaks in the second or third decade, and middle-aged adults begin to loss bone mass at a rate of 0.7% per year [9]. In osteoporosis, the overall density of the skeleton decreases, with thinning of the trabeculae and loss of interconnections. The thinning and loss of interconnections lead to microfractures and eventual collapse of the vertebral body.

Osteoporotic bone fractures are responsible for considerable pain, decreased quality of life, lost workdays, and disability. Up to 30% of patients suffering a hip fracture will require long-term nursing home care. Elderly patients can develop pneumonia and blood clots in the leg veins that can travel to the lungs (pulmonary embolism) due to prolonged bed rest after hip fracture. Osteoporosis has been linked with an increased risk of death. Some 20% of women with a hip fracture will die in the subsequent year as an indirect result of the fracture. Moreover, once a person has experienced a spine fracture due to osteoporosis, he or she is at very high risk of suffering another such fracture in the near future (next few years). About 20% of post-menopausal women who experience a vertebral fracture will suffer a new vertebral fracture of bone in the following year.

A bone is a dynamic, living tissue; its shape and structure continuously evolve throughout life. It has the ability to change its structure by the removal of old bone and replacement with newly formed bone, in a local process called bone remodeling. Osteoporosis is the result of dysfunction of bone remodeling, in that the rate of bone loss is greater than that of bone gain, which gives rise to a porous bone that breaks easily.

Osteoporosis is quite common in elderly people. It is estimated that 25% of women older than 65 may have sustained a crush fracture as a result of osteoporosis [8]. The high prevalence and morbidity associated with osteoporosis has induced the medical community to develop means of prevention of the disease and to search for means of reversing bone loss. Prevention and reversal of bone loss demand a thorough understanding of the remodeling process in bone at the cellular level, including hormonal action. Treatment is exemplified by PTH (see Section 1.6) and mechanical and electromagnetic stimulation.

The goal of treatment of osteoporosis is the prevention of bone fracture by reducing bone loss or, preferably, by increasing bone density and strength. Although early detection and timely treatment of osteoporosis can substantially decrease the risk of future fractures, none of the available treatments for osteoporosis are complete cures. In other words, it is difficult to rebuild completely a bone that has been weakened by osteoporosis. Prevention of osteoporosis is therefore just as important as treatment. Osteoporosis treatment and prevention measures entail lifestyle changes, including quitting cigarette smoking, curtailing excessive alcohol intake, exercising regularly, consuming a balanced diet with adequate calcium and vitamin D, and improving bone remodeling processes. In particular, it is interesting to study the treatment of osteoporosis by reducing bone loss and increasing bone density and strength through improving bone remodeling processes.

1.6 Bone Metabolism

Material metabolism is a chemical process by which cells produce the substances and energy needed to sustain life. As part of metabolism, organic compounds are broken down to provide heat and energy in the process called catabolism. Simpler molecules are also used to build more complex compounds like proteins for growth and repair of tissues, as part of anabolism (the constructive phase of metabolism characterized by the conversion of simple substances into the more complex compounds of living matter).

Many metabolic processes are brought about by the action of enzymes. The overall speed at which an organism carries out its metabolic processes is termed its metabolic rate (or, when the organism is at rest, its basal metabolic rate). Birds, for example, have a high metabolic rate, since they are warm-blooded, and their usual method of locomotion, flight, requires large amounts of energy. Accordingly, birds usually need large amounts of high-quality, energy-rich foods such as seeds or meat, which they must eat frequently. As an example of metabolism, bone metabolism is a lifelong process in which mature bone tissue is removed from the skeleton (a process called bone resorption) and a new bone tissue is formed (a process called ossification or new bone formation). These processes also control the reshaping or replacement of bones following injuries like fracture as well as following microdamage that occurs during normal activities. Metabolism also responds to the functional demands of mechanical loading. In the first year of life, almost 100% of the skeleton is replaced. In adults, remodeling proceeds at about 10% per year.

The cells responsible for bone metabolism are known as osteoblasts, which secrete new bone, and osteoclasts, which break down bone. The structure of bones as well as an adequate supply of calcium requires close cooperation between these two types of cells. Complex signaling pathways are used to achieve proper rates of growth and differentiation. These signaling pathways include the action of several hormones, such as PTH, vitamin D, growth hormone, steroids, and calcitonin, as well as several cytokines. In this way the body is able to maintain adequate levels of calcium required for physiological processes.

After appropriate signaling, osteoclasts move to resorb the surface of bone, followed by deposition of bone by osteoblasts. Together, the cells that are responsible for bone remodeling are known as the basic multicellular unit (BMU), and the temporal duration (i.e., life span) of the BMU is referred to as the bone remodeling period [10].

As was mentioned and stated in Kalfas [11], bone metabolism is under constant regulation by a host of hormonal and local factors. Three of the calcitropic hormones that most affect bone metabolism are parathyroid hormone, vitamin D, and calcitonin. Parathyroid hormone increases the flow of calcium into the calcium pool and maintains the body's extracellular calcium levels at a relatively constant level. This hormone can induce cytoskeletal changes in osteoblasts, which are the only bone cells that have parathyroid hormone receptors. Vitamin D stimulates intestinal and renal calcium-binding proteins and facilitates active calcium transport. Calcitonin is secreted by the parafollicular cells of the thyroid gland in response to an acutely rising plasma calcium level. Calcitonin serves to inhibit calcium-dependent cellular metabolic activity.

Bone metabolism is also affected by a series of proteins, or growth factors, released from platelets, macrophages, and fibroblasts. These proteins cause healing bone to vascularize, solidify, incorporate, and function mechanically. They can induce mesenchymal-derived cells such as monocytes and fibroblasts to migrate, proliferate, and differentiate into bone cells. The proteins that enhance bone healing include the BMPs, insulin-like growth factors, transforming growth factors, platelet derived growth factor, and fibroblast growth factor, among others [12].

1.6.1 Parathyroid Hormone (PTH)

As one of the three important calcitropic hormones mentioned earlier, PTH, or parathormone or parathyrin, is secreted by the chief cells of the parathyroid glands as a polypeptide containing 84 amino acids. It is the most important endocrine regulator of calcium and phosphorus concentration in extracellular fluids. This hormone is secreted from cells of the parathyroid glands and finds its major target cells in bones and kidneys. Another hormone, the parathyroid hormone-related protein, binds to the same receptor as a parathyroid hormone and has major effects on development. Like most other protein hormones, the parathyroid hormone is synthesized as a preprohormone. After intracellular processing, the mature hormone is packaged within the Golgi (a Golgi apparatus represents a cytoplasmic organelle involved in the processing and sorting of proteins and lipids) into regulated secretory vesicles and then secreted into blood by exocytosis. A parathyroid hormone is secreted as a linear protein of 84 amino acids.

PTH enhances the release of calcium from the large reservoir contained in the bones [13]. Bone resorption is the normal destruction of bones by osteoclasts, which are indirectly stimulated by PTH. Stimulation is indirect, as osteoclasts do not have a receptor for PTH; rather, PTH binds to osteoblasts, the cells responsible for creating bones. Binding stimulates the osteoblasts to increase their expression of RANKL (receptor activator of nuclear factor kappa-B ligand) and inhibits their expression of osteoprotegerin (OPG). OPG binds to RANKL and blocks it from interacting with RANK, a receptor for RANKL. The binding of RANKL to RANK (facilitated by the decreased amount of OPG) stimulates these osteoclast precursors to fuse, forming new osteoclasts that ultimately enhance bone resorption.

1.6.2 Vitamin D

Vitamin D, or calcitriol, is a steroid hormone that has long been known for its important role in regulating body levels of calcium and phosphorus, and in the mineralization of bone. More recently, it has become clear that receptors for vitamin D are present in a wide variety of cells, and that this hormone has biological effects that extend far beyond the control of mineral metabolism. Vitamin D is carried in the bloodstream to the liver, where it is converted into the prohormone calcidiol. Circulating calcidiol may then be converted into calcitriol, the biologically active form of vitamin D, either in the kidneys or by monocyte–macrophages in the immune system. Calcitriol mediates its biological effects by binding to vitamin D receptors (VDRs), which are principally located in the nuclei of target cells. The binding of calcitriol to the VDR allows the VDR to act as a transcription factor that modulates the gene expression of transport proteins (such as TRPV6 and calbindin), which are involved in calcium absorption in the intestine.

In fact, vitamin D is well known as a hormone involved in mineral metabolism and bone growth. Its most dramatic effect is to facilitate intestinal absorption of calcium, although it also stimulates absorption of phosphate and magnesium ions. In the absence of vitamin D, dietary calcium is not absorbed at all efficiently. Vitamin D stimulates the expression of a number of proteins involved in transporting calcium from the lumen of the intestine, across the epithelial cells, and into blood. The best studied of these calcium transporters is calbindin, an intracellular protein that ferries calcium across the intestinal epithelial cell.

Numerous effects of vitamin D on bone have been demonstrated. As a transcriptional regulator of bone matrix proteins, it induces the expression of osteocalcin and suppresses the synthesis of type I collagen. In cell cultures, vitamin D stimulates differentiation of osteoclasts. However, studies of humans and animals with vitamin D deficiency or mutations in the vitamin D receptor suggest that these effects are perhaps not of major physiologic importance, and that the crucial effect of vitamin D on bones is to provide the proper balance of calcium and phosphorus to support mineralization. It turns out that vitamin D receptors are present in most if not all cells in the body. Additionally, experiments using cultured cells have demonstrated that vitamin D has potent effects on the growth and differentiation of many types of cells. These findings suggest that vitamin D has physiologic effects much broader than just a role in mineral homeostasis and bone function.

1.6.3 Calcitonin

Calcitonin is a hormone known to participate in calcium and phosphorus metabolism. In mammals, the major source of calcitonin is from the parafollicular cells in the thyroid gland, but it is also synthesized in a wide variety of other tissues, including the lung and intestinal tract. In birds, fish, and amphibians, calcitonin is secreted from the ultimobranchial glands. Calcitonin is a 32 amino acid linear polypeptide hormone produced in humans primarily by the parafollicular cells (also known as C-cells) of the thyroid and in many other animals in the ultimobranchial body. It contains a single disulfide bond, which causes the amino terminus to assume the shape of a ring. Alternative splicing of the calcitonin pre-mRNA can yield an mRNA encoding calcitonin gene-related peptide; that peptide appears to function in the nervous and vascular systems. The calcitonin receptor has been cloned and shown to be a member of the seven-transmembrane, G protein-coupled receptor family.

Calcitonin is formed by the proteolytic cleavage of a larger prepropeptide, which is the product of the CALC1 gene. The CALC1 gene belongs to a superfamily of related protein hormone precursors including islet amyloid precursor protein, calcitonin gene-related peptide, and the precursor of adrenomedullin. Calcitonin suppresses resorption of bones by inhibiting the activity of osteoclasts, the cell type that "digests" bone matrix, releasing calcium and phosphorus into blood.

1.6.4 Insulin-Like Growth Factor

According to Wikipedia, insulin-like growth factors (IGFs) are proteins with high sequence similarity to insulin. IGFs are part of a complex system that cells use to communicate with their physiologic environment. This complex system (often referred to as the IGF "axis") consists of two cell-surface receptors (IGF1R and IGF2R), two ligands (IGF-1 and IGF-2), and a family of six high-affinity IGF-binding proteins (IGFBP 1–6), as well as associated IGFBP degrading enzymes, referred to collectively as proteases.

Their designation as "insulin-like" originated from experiments in which treatment of serum with antibodies to insulin failed to eliminate all insulin activity; the remaining activity was ultimately ascribed to the IGFs. Due to their growth-promoting activities, they were formerly called somatomedins. There are two principal IGFs: IGF-I and IGF-II. Each of these has a number of variant forms, resulting from use of alternative gene promoters and alternative splicing. Structurally, both IGFs resemble insulin in having two chains (A and B) connected by disulfide bonds. Human IGF-I and IGF-II are, respectively, 70 and 67 amino acids in length.

1.6.5 Transforming Growth Factor

Transforming growth factor (sometimes referred to as tumor growth factor, or TGF) is used to describe two classes of polypeptide growth factors, TGF-α and TGF-β. TGF-α is upregulated in some human cancers. It is produced in macrophages, brain cells, and keratinocytes, and induces epithelial development. TGF-β exists in three known subtypes in humans, TGF-β1, TGF-β2, and TGF-β3. These are upregulated in Marfan's syndrome [14] and some human cancers, and they play crucial roles in tissue regeneration, cell differentiation, embryonic development, and regulation of the immune system. Isoforms of TGF-β1 are also thought to be involved in the pathogenesis of preeclampsia (a medical condition in which hypertension arises in pregnancy in association with significant amounts of protein in the urine).

The TGF-β superfamily of cytokines includes the structurally related subfamily of TGF-betas; the subfamily of bone morphogenic proteins (BMPs), decapentaplegic (dpp) and Vg1; the subfamily of Mullerian inhibitory substances; and the subfamily of activins and inhibins. BMP induces ectopic bone formation and plays an important role in the development of the viscera. Ligand binding to its receptor induces the formation of a complex with the type II BMP receptor phosphorylates and activates the type I BMP receptor. The type I BMP receptor then propagates the signal by phosphorylating a family of signal transducers, the SMAD proteins. Currently, eight SMAD proteins have been cloned (SMAD1–7 and SMAD9). Upon phosphorylation by BMP type I receptor, SMAD1 can interact with either SMAD4 or SMAD6. The SMAD1–SMAD6 complex is inactive; the SMAD1–SMAD4 complex, however, triggers the expression of BMP-responsive genes. The ratio between SMAD4 and SMAD6 in cells can modulate the strength of the signal transduced by BMP.

1.6.6 Platelet-Derived Growth Factor

According to Wikipedia, PDGF is one of the numerous growth factors—or proteins that regulate cell growth and division—in bone metabolism. In particular, PDGF plays a significant role in blood vessel formation (angiogenesis), the growth of blood vessels from already existing blood vessel tissue. Uncontrolled angiogenesis is a characteristic of cancer. In chemical terms, platelet-derived growth factor is dimeric glycoprotein composed of two A (-AA) or two B (-BB) chains or a combination of the two (-AB). PDGF is a potent mitogen for cells of mesenchymal origin, including smooth muscle cells and glial cells. In both mouse and human, the PDGF signaling network consists of four ligands, PDGFA–D, and two receptors, PDGFRα and PDGFRβ. All PDGFs function as secreted, disulfide-linked homodimers, but only PDGFA and -B can form functional heterodimers.

PDGFs are mitogenic during early developmental stages, driving the proliferation of undifferentiated mesenchyme and some progenitor populations. During later maturation stages, PDGF signaling has been implicated in tissue remodeling and cellular differentiation, and in inductive events involved in patterning and morphogenesis. In addition to driving mesenchymal proliferation, PDGFs have been shown to direct the migration, differentiation, and function of a variety of specialized mesenchymal and migratory cell types, both during development and in the adult animal [15].

1.6.7 Fibroblast Growth Factor

Fibroblast growth factors, or FGFs, are a family of growth factors involved in angiogenesis, wound healing, and embryonic development. The FGFs are heparin-binding proteins, and their interactions with cell-surface associated heparan sulfate proteoglycans have been shown to be essential for FGF signal transduction. FGFs are key players in the processes of proliferation and differentiation of a wide variety of cells and tissues. The function of FGFs is not restricted to cell growth. Although some of the FGFs do, indeed, induce fibroblast proliferation, the original FGF molecule (FGF-2 or FGF basic) is now known also to induce proliferation of endothelial cells, chondrocytes, smooth muscle cells, and melanocytes, as well as other cells [16]. It can also promote adipocyte differentiation, induce macrophage and fibroblast IL-6 production, stimulate astrocyte migration, and prolong neuronal survival. Thus, the FGF designation is clearly limited by its description as one target cell and one implied biological activity.

1.7 Introduction to Bone Remodeling

Bone remodeling is a complex process performed by the coordinated activities of osteoblasts and osteoclasts. Together, these cells form temporary anatomical structures, called BMUs, which execute bone remodeling. The interactions between osteoblasts and osteoclasts, which guarantee a proper balance between bone gain and loss, is known as coupling [6].

Specific regions of bones are targeted for remodeling due to structural microdamage, thus maintaining the mechanical strength of the skeleton (targeted remodeling) [17,18]. Furthermore, bone remodeling plays a major role in mineral homeostasis by providing access to stores of calcium and phosphate [19]. In this case, bone remodeling occurs at random locations (random remodeling), so every part of the skeleton is remodeled periodically [17,18].

The net amount of old bone removed and new bone restored in the remodeling cycle is a quantity called the bone balance. While coupling is rarely affected, bone balance can vary quite widely in many disease states; for example, in osteoporotic patients, resorption and formation are coupled but there is a negative bone balance (i.e., more bone is resorbed than is replaced by the typical BMUs).

BMUs are constantly remodeling bone tissue in the growing, adult, and senescent skeleton. Most metabolic bone diseases appear when a biochemical or cellular link of this finely organized network is chronically disrupted, such as in osteoporosis, hyperparathyroidism, osteomalacia, and corticosteroid-induced osteopenia. Remodeling also permits the restoration of microdamage caused by fatigue and shock. This constant care of the bone matrix prevents its premature deterioration and maintains its overall strength.

As a major reservoir of body calcium, bone is under the hormonal control of PTH [8]. PTH is the most important hormone regulating calcium homeostasis and bone remodeling. Moreover, PTH is currently involved in numerous clinical trials as an anabolic agent for the treatment of low bone mass in osteoporosis [20]. Interestingly, the overall effect of PTH on bone mass depends primarily on its mode of administration; whereas a continuous increase in PTH levels decreases bone mass, intermittent PTH administration increases bone mass [21–25]. Despite many attempts to identify the source of these differential dosing effects on bone turnover, the precise mechanism remains elusive. Prevention and reversal of bone loss require a thorough understanding of the remodeling process in bones and of the mechanism of bone formation and resorption, including the actions of hormones such as PTH.

Long-term physical activity on a regular basis plays a particularly important role in maintaining healthy bones. Exercise can maintain and increase bone strength by increasing bone mass or by changing bone structures at micro and macro levels. Two main types of exercise are beneficial to bone health: weight-bearing exercise and resistance exercise (lifting weights with arms or legs).

Weight-bearing exercise involves any exercise that is performed while a person is standing so that gravity is exerting a force. Examples of weight-bearing activities are jogging, walking, tennis, dancing, golf, and netball. Activities that are high impact, such as aerobics, running, and jumping, have a greater effect on bone strength than low-impact activities, such as walking and cycling.

Resistance exercises, also called strength training, can also have a positive effect on the health of bones. In moving a heavy weight, the strong muscle contractions place stress or strain on the bone to which the muscles are attached. When a bone is repeatedly strained (as also happens in regular exercise training), it responds by increasing bone mass to become stronger.

Generally speaking, physical activities place mechanical loading on the skeleton and the loading has profound influences on bone remodeling. Disused or reduced loading due to long-term bed rest, cast immobilization, or microgravity conditions (such as that experienced by astronauts in a space station or shuttle) induces obvious bone loss and mineral changes. Conversely, overuse or increased loading has the opposite effect. In tennis players, the bones of their racquet arms display significantly higher bone mineral density and cortical bone content than the bones of their nonplaying arms. Although a great deal of general knowledge exists about the role of mechanical loading on bone remodeling, further understanding of the underlying mechanism is needed.

Pulsed electromagnetic fields (EMFs) have been widely used for the treatment of non-united fractures and congenital pseudarthrosis. Several electrical stimulation systems, such as air-cored and iron-cored coils and solenoids, have been used around the world and claimed to be effective. Electrical parameters such as pulse shape, magnitude, and frequency differ widely, and the exact bone-healing mechanism is still not clearly understood. The application of EMF devices to stimulate osteogenesis is based on the idea of exciting the natural endogenous streaming potentials in bones. In the beginning, an electric current is applied directly onto diseased bone via electrodes, and then a periodic wave is produced by forcing electric current through a wire coil placed over the fracture. Periodic changes of the current then produce the required EMF in bone through Faraday induction.

The most effective medical devices used today are time-dependent EMF, especially pulsed EMF (PEMF). Their frequency range is 1–100 Hz. The physiological frequency ranges from 8 to 30 Hz, which is caused by natural muscle contraction. The subsequently induced EMF in bone tissue is mostly used in clinical therapies. The osteogenesis effect caused by a PEMF device is of great significance to patients, especially those who have already undergone failed surgical intervention [26].

Based on random and prospective clinical studies, the US Food and Drug Administration (FDA) has approved PEMF as a safe and effective way to treat non-union and osteoporosis diseases [27,28], although the specific molecular mechanism is not yet fully understood.

In summary, the current understanding of bone remodeling is primarily based on experimental results in vivo and in vitro. A great deal of research has been conducted on the interactions of autocrine, paracrine, and endocrine activities of receptors and ligands in bone remodeling; the role of bone cells involved in this process at the cellular and genetic level; and the influence of different stimuli and factors such as mechanical stimuli, PEMF, and PTH on bone formation in bone remodeling. Based on these observations, many hypotheses have been proposed as to the role played by different signaling pathways and the communication between bone cells in bone remodeling. However, due to the complexity of the bone regulation system, which involves numerous factors and interactions, understanding of system behavior is still fragmentary.

Mathematical modeling is a powerful tool for reducing ambiguity as to causes and effects in complex systems. It allows one to test various experimental and theoretical hypotheses that might be difficult (such as time- or money consuming) or impossible to test in vivo or in vitro. The development of a pharmaceutical treatment for bone diseases can also be enhanced by computational system biology that uses mathematical modeling to integrate experimental data into a system-level model, enabling various interactions to be investigated efficiently and methodically [29]. However, only a few mathematical models regarding bone remodeling have thus far been proposed [8,20,30–36].

References

1. Schoppet M., Preissner K. T., Hofbauer L. C. RANK ligand and osteoprotegerin—Paracrine regulators of bone metabolism and vascular function. *Arteriosclerosis Thrombosis and Vascular Biology* 22 (4): 549–553 (2002).
2. Agata H., Asahina I., Yamazaki Y., Uchida M., Shinohara Y., Honda M. J., Kagami H., Ueda M. Effective bone engineering with periosteum-derived cells. *Journal of Dental Research* 86 (1): 79–83 (2007).
3. Jones S. J., Boyde A. Experimental study of changes in osteoblastic shape induced by calcitonin and parathyroid extract in an organ-culture system. *Cell and Tissue Research* 169 (4): 449–465 (1976).
4. Zallone A. Z., Teti A., Primavera M. V. Resorption of vital or devitalized bone by isolated osteoclasts in vitro—The role of lining cells. *Cell and Tissue Research* 235 (3): 561–564 (1984).
5. Chambers T. J., Darby J. A., Fuller K. Mammalian collagenase predisposes bone surfaces to osteoclastic resorption. *Cell and Tissue Research* 241 (3): 671–675 (1985).
6. Rodan G. A., Matin T. J. Role of osteoblasts in hormonal control of bone resorption—A hypothesis. *Calcified Tissue International* 33 (4): 349–351 (1981).
7. McSheehy P. M. J., Chambers T. J. Osteoblastic cells mediate osteoclastic responsiveness to parathyroid hormone. *Endocrinology* 118 (2): 824–828 (1986).
8. Kroll M. Parathyroid hormone temporal effects on bone formation and resorption. *Bulletin of Mathematical Biology* 62 (1): 163–188 (2000).
9. Rosenberg A. Skeletal system and soft tissue tumors. In *Robbins pathologic basis of disease*, 5th ed., ed. Cotran R. S., Kumar V., Robbins S. L. Philadelphia: W. B. Saunders Co., pp. 1219–1222 (1994).
10. Pietrzak W. S. *Musculoskeletal tissue regeneration: Biological materials and methods.* Totowa, NJ: Humana Press (2008).
11. Kalfas I. H. Principles of bone healing. *Neurosurgery Focus* 10 (4): 1–4 (2001).
12. Wozney J. M., Rosen V., Celeste A. J., Mitsock L. M., Whitters M. J., Kriz R. W., Hewick R. M., Wang E. A. Novel regulators of bone formation—Molecular clones and activities. *Science* 242 (4885): 1528–1534 (1988).
13. Poole K., Reeve J. Parathyroid hormone—A bone anabolic and catabolic agent. *Current Opinion in Pharmacology* 5 (6): 612–617 (2005).
14. Matt P., Schoenhoff F., Habashi J., Holm T., Van Erp C., Loch D., Carlson O. D., et al. Circulating transforming growth factor-beta in Marfan syndrome. *Circulation* 120 (6): 526–532 (2009).
15. Hoch R. V., Soriano P. Roles of PDGF in animal development. *Development* 130 (20): 4769–4784 (2003).
16. Burgess W. H., Maciag T. The heparin-binding (fibroblast) growth-factor family of proteins. *Annual Review of Biochemistry* 58:575–606 (1989).

17. Burr D. B., Robling A. G., Turner C. H. Effects of biomechanical stress on bones in animals. *Bone* 30 (5): 781–786 (2002).
18. Parfitt A. M. Targeted and nontargeted bone remodeling: relationship to basic multicellular unit origination and progression. *Bone* 30 (1): 5–7 (2002).
19. Black A. J., Topping J., Durham B., Farquharson R. G., Fraser W. D. A detailed assessment of alterations in bone turnover, calcium homeostasis, and bone density in normal pregnancy. *Journal of Bone and Mineral Research* 15 (3): 557–563 (2000).
20. Lemaire V., Tobin F. L., Greller L. D., Cho C. R., Suva L. J. Modeling the interactions between osteoblast and osteoclast activities in bone remodeling. *Journal of Theoretical Biology* 229 (3): 293–309 (2004).
21. Karaplis A. C., Goltzman D. PTH and PTHrP Effects on the skeleton. *Reviews in Endocrine & Metabolic Disorders* 1 (4): 331–341 (2000).
22. Dobnig, H., Turner R. T. The effects of programmed administration of human parathyroid hormone fragment (1-34) on bone histomorphometry and serum chemistry in rats. *Endocrinology* 138 (11): 4607–4612 (1997).
23. Rubin M. R., Bilezikian J. P. New anabolic therapies in osteoporosis. *Endocrinology & Metabolism Clinics of North America* 32 (1): 285–307 (2003).
24. Locklin R. M., Khosla S., Turner R. T., Riggs B. L. Mediators of the biphasic responses of bone to intermittent and continuously administered parathyroid hormone. *Journal of Cellular Biochemistry* 89 (1): 180–190 (2003).
25. Berg C., Neumeyer K., Kirkpatrick P. Teriparatide. *Nature Reviews Drug Discovery* 2 (4): 257–258 (2003).
26. Gossling, H. R, Bernstein, R. A, Abbott, J. Treatment of ununited tibial fractures: A comparison of surgery and pulsed electromagnetic fields (PEMF). *Orthopedics* 15 (6): 711–719 (1992).
27. Fitzsimmons, R. J., Gordon S. L., Kronberg J., Ganey T., Pilla A. A. A pulsing electric field (PEF) increases human chondrocyte proliferation through a transduction pathway involving nitric oxide signaling. *Journal of Orthopaedic Research* 26 (6): 854–859 (2008).
28. Chao E. Y. S., Inoue N., Koo T. K. K., Kim Y. H. Biomechanical considerations of fracture treatment and bone quality maintenance in elderly patients and patients with osteoporosis. *Clinical Orthopaedics and Related Research* Aug. (425): 12–25 (2004).
29. Pivonka P., Zimak J., Smith D. W., Gardiner B. S., Dunstan C. R., Sims N. A., Martin T. J., Mundy G. R. Theoretical investigation of the role of the RANK–RANKL–OPG system in bone remodeling. *Journal of Theoretical Biology* 262 (2): 306–316 (2010).
30. Rattanakul C., Lenbury Y., Krishnamara N., Wolwnd D. J. Modeling of bone formation and resorption mediated by parathyroid hormone: response to estrogen/PTH therapy. *Biosystems* 70 (1): 55–72 (2003).
31. Komarova S. V., Smith R. J., Dixon S. J., Sims S. M., Wahl L. M. Mathematical model predicts a critical role for osteoclast autocrine regulation in the control of bone remodeling. *Bone* 33 (2): 206–215 (2003).
32. Komarova S. V. Mathematical model of paracrine interactions between osteoclasts and osteoblasts predicts anabolic action of parathyroid hormone on bone. *Endocrinology* 146 (8): 3589–3595 (2005).
33. Potter L. K., Greller L. D., Cho C. R., Nuttall M. E., Stroup G. B., Suva L. J., Tobin F. L. Response to continuous and pulsatile PTH dosing: A mathematical model for parathyroid hormone receptor kinetics. *Bone* 37 (2): 159–169 (2005).

34. Pivonka P., Zimak J., Smith D. W., Gardiner B. S., Dunstan C. R., Sims N. A., Martin T. J., Mundy G. R. Model structure and control of bone remodeling: A theoretical study. *Bone* 43 (2): 249–263 (2008).

35. Qin Q. H., Ye J. Q. Thermoelectroelastic solutions for internal bone remodeling under axial and transverse loads. *International Journal of Solids and Structures* 41 (9–10): 2447–2460 (2004).

36. Wang Y. N., Qin Q. H., Kalyanasundaram S. A theoretical model for simulating effect of parathyroid hormone on bone metabolism at cellular level. *Molecular & Cellular Biomechanics* 6 (2): 101–112 (2009).

2

Basic Bone Remodeling Theory

2.1 Introduction

Living bone is a remarkable and extremely adaptable substance in the human body. It continuously undergoes processes of growth, reinforcement, and resorption. These processes are collectively termed *remodeling*. The concept of stress- or strain-induced bone remodeling was first proposed by the German anatomist Julius Wolff [1]. In 1892, Wolff found that the orientation of trabecular bone coincides with the direction of the stress trajectories. He proposed that bone loading is somehow sensed and that the bone adapts its structure accordingly. This principle of functional adaptation is often known as Wolff's law. It occurs in conditions of disuse, such as during immobility, space flight, and long-term bed rest, when bone is lost [2,3], and in overloading, which causes a gain in bone mass [4].

Wolff's law was first described in vitro in 1939 by Glucksmann [5]. Bassett and Becker [6], Shamos, Shamos, and Lavine [7], Justus [8], Cowin [9], and Qin and Ye [10] have proposed various mechanisms for bone remodeling in terms of certain thermal, mechanical, electrical, and chemical properties of bone. In this chapter, we begin with a brief description of adaptive elastic theory [9], followed by a discussion of two typical types of bone remodeling: internal and surface remodeling. A simple theory and its solution related to these two kinds of bone remodeling are presented.

2.2 Adaptive Elastic Theory

An adaptive elastic theory developed by Cowin [9,11] for analyzing the adaptive behavior of living bone under thermal and mechanical loading is briefly reviewed in this section.

As indicated in Cowin [9], certain natural living materials can adjust to their ambient environmentally applied thermal, mechanical, and electrical loads by slowly changing their overall shape and their local mass density or microstructure. These materials include living bones, living wood, and certain saturated porous geological materials. The ability of a bone to adapt to mechanical loads is brought about by continuous bone resorption and bone formation. If these processes occur at different locations, the morphology of the bone is altered. This process was defined as "modeling" by Frost [12]. In a homeostatic equilibrium, resorption and formation are balanced. In that case, old bone is continuously replaced by a new tissue. This ensures that the mechanical integrity of the bone is maintained without any global change in morphology. Frost defined this as "remodeling" [13].

To model these processes mathematically, Cowin and Hegedus [14], Hegedus and Cowin [15], and Cowin and Nachlinger [16] developed a theory of adaptive elasticity for modeling these complex stress or strain adaptation processes with a simple continuum model. The theoretical model is composed of a porous anisotropic linear elastic solid perfused with and surrounded by a fluid. The chemical reactions of the stress (or strain) adaptation process are modeled by the transfer of mass from the fluid to the porous solid matrix, and vice versa. As a result of the chemical reactions, mass is transferred to (from) the solid matrix so that it increases (decreases) either the overall size of the body or the mass density of the body. The rate of change of the overall size and shape of the body is controlled by the surface strain, and the rate of change of mass density at a point is controlled by the local matrix strain. Details of this adaptive model are described in the following subsections.

2.2.1 Two Kinds of Bone Remodeling

Remodeling is fundamental to bone biology. It is a two-stage process carried out by teams of BMUs. Resorption of a packet of bones by osteoclasts is followed by refilling of the resorption cavity by osteoblasts. This sequence typically requires 3–4 months to complete at each locus, and the resorption and refilling cavities, while individually small, may collectively add substantial temporary porosity or "remodeling space" to the bone. If the purpose of elevated remodeling is to remove bone mass, the remodeling space is inconsequential, but if the goal is damage removal or tissue rearrangement, the porosity involved in remodeling can weaken the bone structure. It is therefore essential that remodeling be understood not simply as a fundamental biological process, but also in the context of the load-bearing function of bone.

Generally, there are two major theories of bone remodeling: internal and surface [17]. The distinction between them is as follows. Internal remodeling refers to the resorption or reinforcement of the bone tissue internally by changing the bulk density of the tissue. In a cortical bone, internal remodeling occurs via changes in the diameter of the lamina of the osteons and by the total replacement of osteons. Surface (external) remodeling refers to the resorption or deposition of bone material on the external surface of the bone.

The details of the process of deposition of new lamina on the surface of a bone were described by Currey [18].

Surface remodeling can be induced in the leg bones of animals by superposing axial and/or bending loads. Woo et al. [19] showed that increased physical activity in pigs can cause the periosteal surface of the leg bone to move out and the endosteal surface to move in. Surface remodeling can also be induced in the leg bones of animals by reducing the loads on the limb. In two studies, Uhthoff and Jaworski [20] and Jaworski, Liskovakiar, and Uhthoff [21] immobilized one of the forelimbs of beagles. In the study of Uhthoff and Jaworski [20], young beagles were used and it was found that the endosteal surface showed little movement, whereas there was much resorption on the periosteal surface.

As mentioned in Cowin [11], the theories of both internal and surface remodeling use a simple two-constituent model for bone tissues. The bone matrix—that is, the solid extracellular material and the bone cells—is modeled as a solid structure with interconnected pores, a porous solid. Since bone is adaptively modeled as a linear anisotropic elastic solid, it is assumed that the bone matrix can be modeled as a porous anisotropic linear elastic solid. The extracellular fluid and blood are modeled as a single fluid. Thus, the basic model of bone is a porous, anisotropic linear elastic solid perfused with a fluid.

The chemical reactions that convert body fluids into solid porous bone matrices and vice versa are mediated by the bone cells. Mass, momentum, energy, and entropy are transferred to or from the solid porous bone matrix as a result of the chemical reactions. The rates of these chemical reactions depend on strain and are very slow. The distinction between the two theories lies in the locations at which the chemical reactions occur and the way in which mass is added or removed from the solid porous bone matrix. In the theory of surface remodeling, the chemical reactions occur on the external surfaces only of the bone, and mass is added to or removed from the bone by changing the external shape of the bone. During surface remodeling, the interior of the bone remains at a constant bulk density. In the theory of internal remodeling, the chemical reaction occurs everywhere in the solid porous bone matrix, and mass is added by changing the bulk density of the bone matrix without changing its exterior dimensions. In both cases, the rate and direction of the chemical reaction at a point are determined by the strain at that point.

Cowin [11] mentioned that the surface remodeling theory acknowledges the observed fact that external changes in bone shape are induced by changes in the loading environment of the bone. This theory postulates a causal relationship between the rate of surface deposition or resorption and the strain on the surface of the bone. Bone is considered an open system with regard to mass transport, and the mass of the bone will vary as the external shape of the bone varies. (The theory of external remodeling is discussed in detail in Section 2.3.) In contrast, the theory of internal remodeling acknowledges the orthopedic principle that prolonged straining of a bone tends to make the bone stiffer and denser, whereas prolonged bed rest or inactivity will tend to make the bone less stiff and less dense. That theory postulates a causal relationship between the rate of deposition or resorption of the bone matrix at any point and the strain at that point in the bone matrix. (The theory of internal remodeling is discussed in detail in Section 2.4.)

2.2.2 Surface Bone Remodeling

The model of surface bone remodeling presented in Cowin [11] is described here. In that model, a bone is assumed to be a linear elastic body whose free surfaces move according to an additional specific constitutive relation. The additional constitutive relation for the movement of the free surface is the result of a postulate that the rate of surface deposition or resorption is proportional to the change in the strain in the surface from a reference value of strain. At the reference value of strain there is no movement of the surface.

To express the constitutive equation for the surface movement in equation form, Cowin [11] introduced the following notations: Q is a surface point on the bone (see Figure 2.1), $\mathbf{n}$ is an outward unit normal vector of the tangent plane to the surface of the bone at Q, and U is the velocity of the remodeling surface in the $\mathbf{n}$ direction. The velocity of the surface in any direction in the tangent plane is zero because the surface is not moving tangentially with respect to the bone. The hypothesis for surface remodeling is that the speed of the remodeling surface is linearly proportional to the strain tensor ε_{ij}:

$$U = C_{ij}(Q)[\varepsilon_{ij}(Q) - \varepsilon_{ij}^0(Q)] \tag{2.1}$$

where $\varepsilon_{ij}^0(Q)$ are the reference values of strain at point Q, where no remodeling occurs, and $C_{ij}(\mathbf{n}, Q)$ are the coefficients of the surface remodeling rate that are, in general, dependent on the point Q and the normal $\mathbf{n}$ to the surface at Q.

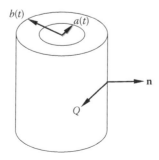

FIGURE 2.1
Geometry of the diaphysis of the bone.

The surface remodeling rate coefficients and the reference values of strain are phenomenological coefficients of the bone surface and can only be determined experimentally. To simplify the analysis, Cowin [11] assumed that the surface remodeling rate coefficients C_{ij} are not site dependent; that is, they are independent of the position of the surface point Q. He also postulated that C_{ij} are independent of strain. Equation (2.1) gives the normal velocity of the surface at Q as a function of the existing strain state at the point Q. If the right-hand side of Equation (2.1) is positive, the surface will be growing by the deposition of material. If the right-hand side of Equation (2.1) is negative, the surface will be resorbing.

It should be mentioned that Equation (2.1) by itself does not constitute the complete theory. The theory is completed by assuming that bone is a linearly elastic material. Thus, the complete theory is a modification of linear elasticity in which the external surfaces of the bone move according to the rule prescribed by Equation (2.1). A boundary value problem can be formulated in the same manner as a boundary value problem in linear elastostatics, but it is necessary to specify the boundary conditions for a specific time period. As the bone evolves to a new shape, the stress and strain states will be varying quasistatically. At any instant, the bone will behave exactly as an elastic body, but moving boundaries will cause local stress and strain to redistribute themselves slowly with time.

2.2.3 Internal Bone Remodeling

As described in Cowin [11], the small theory of internal bone remodeling is an adaptation of the theory of equilibrium of elastic bodies. The theory models the bone matrix as a chemically reacting porous elastic solid. The bulk density ρ of the porous solid is written as the product of γ and v:

$$\rho = \gamma v \tag{2.2}$$

where γ is the density of the material that composes the matrix structure, and v is the volume fraction of that material.

Both γ and v can be considered field variables, by the same arguments as employed by Goodman and Cowin [22]. Let ξ denote the value of the volume fraction v of the matrix material in an unstrained reference state. The density γ of the material composing the matrix is assumed to be a constant; hence, the conservation of mass will give the equation governing ξ. It is also assumed that there exists a unique zero-strain reference state for all values of ξ. Thus, ξ may change without changing the reference state for strain. The change in the volume fraction from a reference volume fraction ξ_0 is denoted by e:

$$e = \xi - \xi_0 \tag{2.3}$$

The basic kinematic variables and also the independent variables for the theory of internal bone remodeling are the six components of the strain matrix ε_{ij} and the change in volume fraction e of the matrix material from a reference value ξ_0.

The equations of the governing system for this theory are [15]

$$\varepsilon_{ij} = (u_{i,j} + u_{j,i})/2 \tag{2.4}$$

$$\sigma_{ij,j} + \gamma(\xi_0 + e)b_i = 0 \tag{2.5}$$

$$\sigma_{ij} = (\xi_0 + e)C_{ijkm}(e)\varepsilon_{km} \tag{2.6}$$

$$\dot{e} = A(e) + A_{km}(e)\varepsilon_{km} \tag{2.7}$$

where u_i are displacement components, σ_{ij} the components of the stress tensor, b_i the components of the body force, $C_{ijkm}(e)$ components of the fourth-rank elasticity tensor, $A(e)$ and $A_{ij}(e)$ the material coefficients dependent on the change in volume fraction e of the adaptive elastic material from the reference volume fraction, and the superimposed dot indicates the material time derivative.

Equation (2.4) is the strain–displacement relations for small strain; Equation (2.5) represents the condition of equilibrium in terms of stress. Equation (2.6) stands for a generalization of Hooke's law; Equation (2.7) represents the remodeling rate equation, and it specifies the rate of change of the volume fraction as a function of ξ and ε_{ij}. A positive value of $\dot{e}$ means that the volume fraction of elastic material in increasing, whereas a negative value means that the volume fraction is decreasing. Equation (2.7) is obtained from the conservation of mass and the constitutive assumption that the rate of mass deposition or absorption is dependent on only the volume fraction e and ε_{ij}. The system of Equations (2.6) and (2.7) is an elementary mathematical model of Wolff's law. Equation (2.6) is a statement that the moduli occurring in Hooke's law actually depend on the volume fraction of solid matrix material present. Equation (2.7) is an evolutionary law for the volume fraction of matrix material.

The theory just described involves the functions $A(e)$, $A_{ij}(e)$, and $C_{ijkm}(e)$, characterizing the material properties. Cowin [11] mentioned that no data exist in the literature on the values of the functions $A(e)$ and $A_{ij}(e)$, and the data on $C_{ijkm}(e)$ suggest that it can be approximated as a linear function of e. Hegedus and Cowin [15] introduced an approximation scheme that gave $C_{ijkm}(e)$ as a linear function of e. This scheme involved a series expansion in which terms of the orders e^3, $|\varepsilon_{ij}|e^2$, and $|\varepsilon_{ij}|^2 e$ were ignored and terms of the orders e, $|\varepsilon_{ij}|$, $|\varepsilon_{ij}|e$, and e^2 retained. The scheme showed that $A_{ij}(e)$ were also linear in e, whereas $A(e)$ was quadratic in e, and thus the constitutive relations (2.6) and (2.7) were approximated by

$$\sigma_{ij} = (\xi_0 C_{ijkm}^0 + eC_{ijkm}^1)\varepsilon_{km} \tag{2.8}$$

and

$$\dot{e} = c_0 + c_1 e + c_2 e^2 + A_{ij}^0 \varepsilon_{ij} + e A_{ij}^1 \varepsilon_{ij} \tag{2.9}$$

where C_{ijkm}^0, C_{ijkm}^1, c_0, c_1, c_2, A_{ij}^0, and A_{ij}^1 are material constants. Equations (2.1)–(2.9) consist of the basic formulation of surface and internal bone remodeling. More details as to the solution procedure and application of bone remodeling theory are described in the remaining sections of this chapter.

2.3 A Simple Theory of Surface Bone Remodeling

In the previous section a simple constitutive equation (2.1) for surface bone remodeling was presented. Based on Equation (2.1) and the theory of linear elastic solid material, Cowin and Van Buskirk [23] developed a simple theory of surface bone remodeling and applied it to the problem of predicting the surface remodeling that would occur in the diaphysial region of a long bone as a result of superposed compressive load and as a result of a force-fitted medullary pin. Papathanasopoulou, Fotiadis, and Massalas [24] extended it to the case of poroelastic material with fluid. A brief review of the development in Cowin and Van Buskirk and Papathanasopoulou et al. is presented in this section.

2.3.1 Basic Equations of the Theory

As mentioned in Subsection 2.2.2 and also indicated in Cowin and Van Buskirk [23], Equation (2.1) by itself does not constitute the complete theory. The theory of surface bone remodeling is completed by considering a bone as a linear elastic solid and implementing the related elastic equations. For simplicity, Cowin and Van Buskirk assumed the bone to be transversely isotropic with the axis of a long bone as the axis of symmetry and considered to have only cylindrical boundaries (see Figure 2.1). Therefore, the basic equations can be written within the framework of a cylindrical coordinate system. Based on that assumption, the strain–stress relations for the bone are written as

$$\varepsilon_{rr} = \frac{1}{E_T}(\sigma_{rr} - \mu_T \sigma_{\theta\theta}) - \frac{\mu_A}{E_A}\sigma_{zz}, \quad \varepsilon_{r\theta} = \frac{1}{2G_T}\sigma_{r\theta},$$

$$\varepsilon_{\theta\theta} = \frac{1}{E_T}(\sigma_{\theta\theta} - \mu_T \sigma_{rr}) - \frac{\mu_A}{E_A}\sigma_{zz}, \quad \varepsilon_{\theta z} = \frac{1}{2G_A}\sigma_{\theta z}, \tag{2.10}$$

$$\varepsilon_{zz} = \frac{\mu_A}{E_A}(\sigma_{\theta\theta} + \sigma_{rr}) + \frac{1}{E_A}\sigma_{zz}, \quad \varepsilon_{rz} = \frac{1}{2G_A}\sigma_{rz}$$

with the subscripts A and T for the axial and transverse directions, respectively; E_A and E_T for the Young's moduli; G_A and G_T for the moduli of rigidity; and μ_A and μ_T for the Poisson's ratios. The inverse of Equation (2.10) gives the stiffness equations as

$$\sigma_{rr} = (\lambda_2 + 2G_T)\varepsilon_{rr} + \lambda_2\varepsilon_{\theta\theta} + \lambda_1\varepsilon_{zz}, \quad \sigma_{r\theta} = 2G_T\varepsilon_{r\theta},$$

$$\sigma_{\theta\theta} = (\lambda_2 + 2G_T)\varepsilon_{\theta\theta} + \lambda_2\varepsilon_{rr} + \lambda_1\varepsilon_{zz}, \quad \sigma_{\theta z} = 2G_A\varepsilon_{\theta z}, \tag{2.11}$$

$$\sigma_{zz} = \lambda_1(\varepsilon_{rr} + \varepsilon_{\theta\theta}) + (\lambda_1 + 2G_1)\varepsilon_{zz}, \quad \sigma_{rz} = 2G_A\varepsilon_{rz}$$

where

$$\lambda_1 = \frac{\mu_A E_A E_T}{(1-\mu_T)E_A - 2\mu_A^2 E_T},$$

$$\lambda_2 = \frac{\mu_T E_A E_T + \mu_A^2 E_T^2}{(1-\mu_T^2)E_A - 2\mu_A E_T(1+\mu_T)}, \tag{2.12}$$

$$G_1 = \frac{(1-\mu_T)E_A^2 - \mu_A E_A E_T}{(1-\mu_T)E_A - 2\mu_A^2 E_T}$$

The strain–displacement relations in cylindrical coordinates are

$$\varepsilon_{rr} = \frac{\partial u}{\partial r}, \quad \varepsilon_{\theta\theta} = \frac{\partial v}{r\,\partial\theta} + \frac{u}{r}, \quad \varepsilon_{zz} = \frac{\partial w}{\partial z},$$

$$\varepsilon_{r\theta} = \frac{1}{2}\left(\frac{\partial u}{r\,\partial\theta} + \frac{\partial v}{\partial r} - \frac{v}{r}\right), \quad \varepsilon_{rz} = \frac{1}{2}\left(\frac{\partial u}{\partial z} + \frac{\partial w}{\partial r}\right), \tag{2.13}$$

$$\varepsilon_{\theta z} = \frac{1}{2}\left(\frac{\partial w}{r\,\partial\theta} + \frac{\partial v}{\partial z}\right)$$

where u, v, and w are the components of the displacement vector in the radial, tangential, and axial directions, respectively. In the absence of body forces, the corresponding equilibrium equations are

$$\frac{\partial\sigma_{rr}}{\partial r} + \frac{1}{r}\frac{\partial\sigma_{r\theta}}{\partial\theta} + \frac{\partial\sigma_{rz}}{\partial z} + \frac{\sigma_{rr} - \sigma_{\theta\theta}}{r} = 0,$$

$$\frac{\partial\sigma_{r\theta}}{\partial r} + \frac{1}{r}\frac{\partial\sigma_{\theta\theta}}{\partial\theta} + \frac{\partial\sigma_{\theta z}}{\partial z} + \frac{2}{r}\sigma_{r\theta} = 0, \tag{2.14}$$

$$\frac{\partial\sigma_{rz}}{\partial r} + \frac{1}{r}\frac{\partial\sigma_{\theta z}}{\partial\theta} + \frac{\partial\sigma_{zz}}{\partial z} + \frac{1}{r}\sigma_{rz} = 0$$

The surface remodeling equation (2.1) can be written in cylindrical coordinates as

$$U = C_R \varepsilon_{rr} + C_Z \varepsilon_{zz} + C_\theta \varepsilon_{\theta\theta} + C_{RZ} \varepsilon_{rz} + C_{R\theta} \varepsilon_{r\theta} + C_{\theta Z} \varepsilon_{\theta z} - C^0 \tag{2.15}$$

where

$$C^0 = C_R \varepsilon_{rr}^0 + C_Z \varepsilon_{zz}^0 + C_\theta \varepsilon_{\theta\theta}^0 + C_{RZ} \varepsilon_{rz}^0 + C_{R\theta} \varepsilon_{r\theta}^0 + C_{\theta Z} \varepsilon_{\theta z}^0 \tag{2.16}$$

with the subscripts R, Z, and θ referring to the radial, axial, and tangential directions, respectively. Substituting Equation (2.10) into Equation (2.15), we have

$$U = B_R \sigma_{rr} + B_Z \sigma_{zz} + B_\theta \sigma_{\theta\theta} + B_{RZ} \sigma_{rz} + B_{R\theta} \sigma_{r\theta} + B_{\theta Z} \sigma_{\theta z} - C^0 \tag{2.17}$$

where

$$
\begin{aligned}
B_R &= \frac{C_R}{E_R} - \frac{\mu_Z C_Z}{E_Z} - \frac{\mu_R C_\theta}{E_R}, & B_{R\theta} &= \frac{C_{R\theta}}{2G_R}, \\[2mm]
B_\theta &= \frac{C_\theta}{E_R} - \frac{\mu_Z C_Z}{E_Z} - \frac{\mu_R C_R}{E_R}, & B_{RZ} &= \frac{C_{RZ}}{2G_Z}, \\[2mm]
B_Z &= \frac{C_Z}{E_Z} - (C_R + C_\theta)\frac{\mu_Z}{E_Z}, & B_{RZ} &= \frac{C_{RZ}}{2G_Z}
\end{aligned}
\tag{2.18}
$$

Equations (2.10)–(2.18) are the basic equations for the solution of bone surface remodeling problems.

2.3.2 Bone Remodeling of Diaphysial Surfaces

To illustrate the application of the theory just described, a simple example of surface remodeling occurring in a diaphysial region of a long bone under a constant load [23] is summarized in this subsection.

Cowin and Van Buskirk [23] considered a hollow circular cylinder composed of linearly elastic bone materials subjected to a constant compressive axial load P for an indefinite time period. As a result of the applied load P, the hollow cylinder, which initially had an internal radius a_o and an external radius b_o, will remodel into a cylinder with an internal radius a_∞ and an external radius b_∞. The objective of the theory is to predict the instantaneous values $a(t)$ and $b(t)$ of these radii. Cowin and Van Buskirk assumed that the surface remodeling rate coefficients are different on the periosteal and endosteal surfaces, but that each is constant on these surfaces throughout the diaphysis. Since σ_{zz} is the only nonzero stress, it follows from Equation (2.17) that

$$U_p = B_{Zp} \sigma_{zz} - C_p^0 \tag{2.19}$$

on the periosteal surface and

$$U_e = B_{Ze}\sigma_{zz} - C_e^0 \tag{2.20}$$

on the endosteal surface, where the subscripts p and e refer to periosteal and endosteal, respectively. Since U_e and U_p are the velocities normal to the inner and outer surfaces of the cylinders, respectively, they are calculated as

$$U_e = -\frac{da}{at}, \quad U_p = \frac{db}{at} \tag{2.21}$$

where the minus sign appearing in the expression for U_e denotes that the outward normal of the endosteal surface is in the negative coordinate direction. The stress in the axial direction of the hollow cylinder is given by

$$\sigma_{zz} = \frac{-P}{\pi(b^2 - a^2)} \tag{2.22}$$

Substituting Equations (2.21) and (2.22) into Equations (2.19) and (2.20), we have

$$\frac{1}{B_{Zp}}\frac{db}{dt} = \frac{-P}{\pi(b^2 - a^2)} - \frac{C_p^0}{B_{Zp}} \tag{2.23}$$

$$\frac{-1}{B_{Ze}}\frac{da}{dt} = \frac{-P}{\pi(b^2 - a^2)} - \frac{C_e^0}{B_{Ze}} \tag{2.24}$$

Subtracting Equation (2.24) from Equation (2.23), we have the differential equation

$$\frac{1}{B_{Ze}}\frac{da}{dt} + \frac{1}{B_{Zp}}\frac{db}{dt} = \frac{C_e^0}{B_{Ze}} - \frac{C_p^0}{B_{Zp}} \tag{2.25}$$

which has the solution

$$\frac{a(t)}{B_{Ze}} + \frac{b(t)}{B_{Zp}} = \left(\frac{C_e^0}{B_{Ze}} - \frac{C_p^0}{B_{Zp}}\right)t + \frac{a_0}{B_{Ze}} + \frac{b_0}{B_{Zp}} \tag{2.26}$$

where the initial size of the hollow cylinder has been used as an initial condition. So that the size of the cylinder does not become unbounded in time, it is necessary for the first term on the right-hand side of Equation (2.26) to vanish; thus, a constant, say σ_0, can be defined as

$$\sigma_0 = \frac{C_e^0}{B_{Ze}} = \frac{C_p^0}{B_{Zp}} \tag{2.27}$$

To obtain an explicit and readily understood answer, Cowin and Van Buskirk assumed that changes in the radii are small in terms of initial radius a_0. Specifically, they assumed that the squares in the quantities ε and η, where

$$\varepsilon = \frac{a}{a_0} - 1, \quad \eta = \frac{b}{b_0} - 1 \tag{2.28}$$

are negligible. The definitions (2.28) and the definition (2.27) are substituted into Equation (2.26), leading to

$$\varepsilon(t)\frac{a_0}{B_{Ze}} + \eta(t)\frac{b_0}{B_{Zp}} = 0 \tag{2.29}$$

When Equations (2.28), (2.29), and the assumed smallness in the change of radii are employed in Equation (2.23), it takes the form

$$b_0 \frac{d\eta}{dt} = B_{Zp}\alpha - \beta b_0 \eta \tag{2.30}$$

where

$$\alpha = \frac{P}{\pi(b_0^2 - a_0^2)} + \sigma_0 \tag{2.31}$$

$$\beta = \frac{-2P(B_{Ze}a_0 + B_{Zp}b_0)}{\pi(b_0^2 - a_0^2)^2} \tag{2.32}$$

The solutions to Equations (2.23) and (2.24) are easily obtained with this approximation as

$$a(t) = a_0 - \frac{\alpha}{\beta} B_{Ze}(1 - e^{-\beta t}), \quad b(t) = b_0 + \frac{\alpha}{\beta} B_{Zp}(1 - e^{-\beta t}) \tag{2.33}$$

The final radii of the cylinder are determined from Equation (2.33) as the limiting values of a and b, as t tends to infinity,

$$a_\infty = a_0 - \frac{\alpha}{\beta} B_{Ze}, \quad b_\infty = b_0 + \frac{\alpha}{\beta} B_{Zp} \tag{2.34}$$

The stress in the axial direction is obtained by substituting Equation (2.33) back into Equation (2.22) as

$$\sigma_{zz} = \frac{-P}{\pi(b_0^2 - a_0^2)}e^{-\beta t} + \sigma_0(1 - e^{-\beta t}) \qquad (2.35)$$

The final stress is obtained by taking the value of the stress as t tends to infinity:

$$\sigma_{zz}|_{t \to \infty} = \sigma_0 \qquad (2.36)$$

This value of the stress indicates that the strain has returned to its reference value E_{zz}^0 and that the speed of the surface remodeling has approached zero for both endosteal and periosteal surfaces.

2.3.3 Extension to Poroelastic Bone with Fluid

On the basis of Biot's theory of consolidation, Papathanasopoulou et al. [24] modified the theory described in Section 2.3.1 by incorporating fluid flow in the description of the process of bone remodeling. In the modified theory, bone was assumed to be a porous isotropic solid through the pores of which a viscous compressible fluid flows, and the formulation described in Section 2.3 was revised using the theory of consolidation introduced by Biot [25]. The system of solid plus fluid is assumed to be a system with conservation properties. The solid part is considered to have compressibility and shearing rigidity and the fluid to be compressible. This section presents a brief review of this modified surface bone remodeling theory [24].

Considering the effect of fluid flow, the stress tensor in a porous material can be expressed as

$$\bar{\sigma}_{ij} = \sigma_{ij} + \delta_{ij}\sigma \qquad (2.37)$$

where δ_{ij} are Kronecker's deltas, σ_{ij} are the stress components applied to the solid part, and σ represents the total normal force applied to the fluid part of the faces of a cube of the bulk material with unit size (here, bone matrix plus fluid). Sigma is related to the hydrostatic pressure p of the fluid in the pores by

$$\sigma = -fp \qquad (2.38)$$

where f is the porosity defined as

$$f = V_p/V_b \qquad (2.39)$$

where V_p is the volume of the pores contained in a sample of bulk volume V_b. Thus, f represents the fraction of the volume of the porous material occupied by the pores.

The stress–strain equations (2.11) now become

$$\sigma_{rr} = 2N\varepsilon_{rr} + AE + Q\varepsilon, \quad \sigma_{r\theta} = N\varepsilon_{r\theta},$$

$$\sigma_{\theta\theta} = 2N\varepsilon_{\theta\theta} + AE + Q\varepsilon, \quad \sigma_{\theta z} = N\varepsilon_{\theta z}, \qquad (2.40)$$

$$\sigma_{zz} = 2N\varepsilon_{zz} + AE + Q\varepsilon, \quad \sigma_{rz} = N\varepsilon_{rz}, \quad \sigma = QE + R\varepsilon$$

for an isotropic poroelastic material, where A, N, R, and Q are the elastic constants of the material, in accordance with Biot's formulation [25], and E and ε are the dilatations of the solid and fluid:

$$E = \varepsilon_{rr} + \varepsilon_{\theta\theta} + \varepsilon_{zz} \qquad (2.41)$$

for solid and

$$\varepsilon = \varepsilon_{rr}^{(f)} + \varepsilon_{\theta\theta}^{(f)} + \varepsilon_{zz}^{(f)} \qquad (2.42)$$

for fluid, where the superscript f represents the related variable being associated with fluid.

The inverse of Equation (2.40) gives the strain–stress relations as

$$\varepsilon_{rr} = \frac{1}{2N}\left[(1+q)\sigma_{rr} + q(\sigma_{\theta\theta} + \sigma_{zz}) - \frac{Q}{R}(3q+1)\sigma \right],$$

$$\varepsilon_{\theta\theta} = \frac{1}{2N}\left[(1+q)\sigma_{\theta\theta} + q(\sigma_{rr} + \sigma_{zz}) - \frac{Q}{R}(3q+1)\sigma \right],$$

$$\varepsilon_{zz} = \frac{1}{2N}\left[(1+q)\sigma_{zz} + q(\sigma_{\theta\theta} + \sigma_{rr}) - \frac{Q}{R}(3q+1)\sigma \right], \qquad (2.43)$$

$$\varepsilon_{\theta z} = \sigma_{\theta z}/N, \quad \varepsilon_{rz} = \sigma_{rz}/N, \quad \varepsilon_{r\theta} = \sigma_{r\theta}/N$$

and

$$E = \frac{3q+1}{2N}\left[\sigma_{rr} + \sigma_{\theta\theta} + \sigma_{zz} - \frac{3Q\sigma}{R} \right],$$

$$\varepsilon = -Qs(\sigma_{rr} + \sigma_{\theta\theta} + \sigma_{zz}) + \left(1 + 3Q^2 s\right)\frac{\sigma}{R} \qquad (2.44)$$

where

$$q = \frac{Q^2 - AR}{(2N + 3A) - 3Q^2}, \quad s = \frac{q}{Q^2 - AR} \tag{2.45}$$

The total stress field of the bulk material, in the absence of body forces, satisfies the equilibrium equations (2.14) provided that σ_{rr}, $\sigma_{\theta\theta}$, and σ_{zz} in Equation (2.14) are, respectively, replaced by $(\sigma_{rr} + \sigma)$, $(\sigma_{\theta\theta} + \sigma)$, and $(\sigma_{zz} + \sigma)$; that is,

$$\frac{\partial(\sigma_{rr} + \sigma)}{\partial r} + \frac{1}{r}\frac{\partial \sigma_{r\theta}}{\partial \theta} + \frac{\partial \sigma_{rz}}{\partial z} + \frac{\sigma_{rr} - \sigma_{\theta\theta}}{r} = 0,$$

$$\frac{\partial \sigma_{r\theta}}{\partial r} + \frac{1}{r}\frac{\partial(\sigma_{\theta\theta} + \sigma)}{\partial \theta} + \frac{\partial \sigma_{\theta z}}{\partial z} + \frac{2}{r}\sigma_{r\theta} = 0, \tag{2.46}$$

$$\frac{\partial \sigma_{rz}}{\partial r} + \frac{1}{r}\frac{\partial \sigma_{\theta z}}{\partial \theta} + \frac{\partial(\sigma_{zz} + \sigma)}{\partial z} + \frac{1}{r}\sigma_{rz} = 0$$

Darcy's law governing the flow of a fluid is

$$\frac{\partial \sigma}{\partial r} = C\frac{\partial}{\partial t}(U_r - u_r), \quad \frac{\partial \sigma}{r\,\partial \theta} = C\frac{\partial}{\partial t}(U_\theta - u_\theta), \quad \frac{\partial \sigma}{\partial z} = C\frac{\partial}{\partial t}(U_z - u_z) \tag{2.47}$$

where u_i are the average displacements in the solid, U_i denote the average displacements in the fluid, and C is a constant that depends on the permeability κ_m, the porosity f of the medium, and the viscosity η of the fluid [26]:

$$C = \frac{\eta f^2}{\kappa_m} \tag{2.48}$$

Since the constitutive equation (2.6) is not suitable for the poroelastic material under consideration, Papathanasopoulou et al. rewrote Equation (2.40) as follows:

$$\begin{Bmatrix} \sigma_{rr} + \sigma \\ \sigma_{\theta\theta} + \sigma \\ \sigma_{zz} + \sigma \\ \sigma_{\theta z} \\ \sigma_{zr} \\ \sigma_{r\theta} \end{Bmatrix} = \begin{bmatrix} 2N + X & X & X & 0 & 0 & 0 \\ A + Q & 2N + X & X & 0 & 0 & 0 \\ X & X & 2N + X & 0 & 0 & 0 \\ 0 & 0 & 0 & N & 0 & 0 \\ 0 & 0 & 0 & 0 & N & 0 \\ 0 & 0 & 0 & 0 & 0 & N \end{bmatrix} \begin{Bmatrix} E_{rr} + x_1\varepsilon \\ E_{\theta\theta} + x_2\varepsilon \\ E_{zz} + x_3\varepsilon \\ E_{\theta z} \\ E_{zr} \\ E_{r\theta} \end{Bmatrix} \tag{2.49}$$

where $X = A + Q$. It follows from Equation (2.49) that

$$
\begin{aligned}
(2N+X)x_1 + X(x_2 + x_3) &= Q+R \\
(2N+X)x_2 + X(x_3 + x_1) &= Q+R \quad \Rightarrow x_1 = x_2 = x_3 = \frac{Q+R}{2N+3A+3Q} \\
(2N+X)x_3 + X(x_1 + x_2) &= Q+R
\end{aligned}
\tag{2.50}
$$

where x_1, x_2, and x_3 represent parameters related to the fraction of the fluid dilatation in the r, θ, and z directions, respectively.

Making use of Equations (2.6) and (2.49), the stress–strain relation for isotropic poroelastic bone can be written as

$$
\sigma_{ij} + \delta_{ij}\sigma = C_{ijkm}\left(\varepsilon_{km} + \delta_{km}\frac{Q+R}{2N+3A+3Q}\varepsilon \right)
\tag{2.51}
$$

Using an analogous approach to that of Cowin and Van Buskirk [23] and Equation (2.51), the constitutive equation for the speed of the remodeling surface can be written as

$$
U = C_{ij}(\mathbf{n}, S)\left[\varepsilon_{ij}(S) - \varepsilon_{ij}^0(S) + \delta_{ij}\frac{Q+R}{2N+3A+3Q}\left(\varepsilon(S) - \varepsilon^0(S)\right) \right]
\tag{2.52}
$$

where S is used (instead of Q in the previous section) to represent a surface point, to avoid confusion with constant Q used in this section. In cylindrical coordinates, this expression can be written as

$$
\begin{aligned}
U = C_R\left(\varepsilon_{rr} + \frac{Q+R}{2N+3A+3Q}\varepsilon \right) + C_Z\left(\varepsilon_{zz} + \frac{Q+R}{2N+3A+3Q}\varepsilon \right) \\
+ C_\theta\left(\varepsilon_{\theta\theta} + \frac{Q+R}{2N+3A+3Q}\varepsilon \right) + C_{RZ}\varepsilon_{rz} + C_{R\theta}\varepsilon_{r\theta} + C_{\theta Z}\varepsilon_{\theta z} - C^0
\end{aligned}
\tag{2.53}
$$

where

$$
\begin{aligned}
C^0 = C_R\left(\varepsilon_{rr}^0 + \frac{Q+R}{2N+3A+3Q}\varepsilon^0 \right) + C_Z\left(\varepsilon_{zz}^0 + \frac{Q+R}{2N+3A+3Q}\varepsilon^0 \right) \\
+ C_\theta\left(\varepsilon_{\theta\theta}^0 + \frac{Q+R}{2N+3A+3Q}\varepsilon^0 \right) + C_{RZ}\varepsilon_{rz}^0 + C_{R\theta}\varepsilon_{r\theta}^0 + C_{\theta Z}\varepsilon_{\theta z}^0
\end{aligned}
\tag{2.54}
$$

The constitutive equation for the speed of the remodeling surface can be written in terms of stresses as

$$
U = B_R\sigma_{rr} + B_Z\sigma_{zz} + B_\theta\sigma_{\theta\theta} + B_{RZ}\sigma_{rz} + B_{R\theta}\sigma_{r\theta} + B_{\theta Z}\sigma_{\theta z} + B\sigma - C^0
\tag{2.55}
$$

where

$$B_R = \left[\frac{1+q}{2N} + \frac{(Q+R)Qs}{2N+3A+3Q}\right]C_R + \left[\frac{q}{2N} + \frac{(Q+R)Qs}{2N+3A+3Q}\right](C_\theta + C_Z),$$

$$B_\theta = \left[\frac{1+q}{2N} + \frac{(Q+R)Qs}{2N+3A+3Q}\right]C_\theta + \left[\frac{q}{2N} + \frac{(Q+R)Qs}{2N+3A+3Q}\right](C_R + C_Z),$$

$$B_Z = \left[\frac{1+q}{2N} + \frac{(Q+R)Qs}{2N+3A+3Q}\right]C_Z + \left[\frac{q}{2N} + \frac{(Q+R)Qs}{2N+3A+3Q}\right](C_R + C_\theta), \quad (2.56)$$

$$B_{R\theta} = \frac{C_{R\theta}}{N}, \quad B_{RZ} = \frac{C_{RZ}}{N}, \quad B_{\theta Z} = \frac{C_{\theta Z}}{N},$$

$$B = -\left[\frac{Q(3q+1)}{2NR} + \frac{(Q+R)Qs}{2N+3A+3Q}\left(\frac{1}{R} + \frac{3sQ^2}{R}\right)\right](C_\theta + C_Z + C_R)$$

2.4 A Simple Theory of Internal Bone Remodeling

Based on Equations (2.4)–(2.7), Cowin and Van Buskirk [27] presented a theoretical solution for internal bone remodeling induced by a medullary pin; Tsili [28] derived a solution for the case of bone remodeling induced by casting a broken femur. Papathanasopoulou et al. [29] developed a model for internal remodeling of poroelastic bone filled with fluid. In this section, the developments in Tsili and Papathanasopoulou et al. are briefly summarized for reference in subsequent chapters.

2.4.1 Internal Remodeling Induced by Casting a Broken Femur

Tsili [28] considered a femur in a steady state subjected to a constant tensile axial load $P_0(t)$ due to the weight of the lower leg. The bone is assumed to have a uniform reference volume fraction ξ_0 and e is constant in the whole solution domain. At $t = 0$, the bone was cast. Then this bone was under the tensile load $P_0(t)$ and under a constant external pressure $P_1(t)$ due to the cast. Tsili's purpose with this work was to predict $e(t_0)$, where t_0 is the time at which the plaster cast is removed. The bone (femur) is modeled as a hollow circular cylinder with constant inner and outer radii a and b, respectively. The boundary conditions of this problem are

$$\sigma_{rr} = \sigma_{r\theta} = \sigma_{rz} = 0 \qquad \text{at} \quad r = a \qquad\qquad (2.57)$$

$$\sigma_{rr} = -P_1 \quad \sigma_{r\theta} = \sigma_{rz} = 0 \qquad \text{at} \quad r = b \qquad\qquad (2.58)$$

and

$$\int_s \sigma_{zz}ds = P_0 \quad \Rightarrow \quad \sigma_{zz} = \frac{P_0}{\pi(b^2 - a^2)} \tag{2.59}$$

The remodeling rate equation of this problem is defined by Equation (2.7). The corresponding stress–stress relations, strain-displacement relations, and the equilibrium equation are, respectively, defined by Equations (2.11), (2.13), and (2.14). Tsili assumed the solution to the preceding problem to be in the form

$$u_r = A(t)r + B(t)/r, \quad u_\theta = 0, \quad u_z = C(t)z \tag{2.60}$$

where $A(t)$, $B(t)$, and $C(t)$ are unknown functions to be determined from the boundary conditions (2.57)–(2.59). Substituting Equation (2.60) into Equation (2.13), we have

$$\varepsilon_{rr} = A(t) - B(t) / r^2, \quad \varepsilon_{\theta\theta} = A(t) + B(t) / r^2,$$

$$\varepsilon_{zz} = C(t), \quad \varepsilon_{r\theta} = \varepsilon_{rz} = \varepsilon_{z\theta} = 0 \tag{2.61}$$

The assumed strain (2.61) is substituted into the stress–strain relations (2.11), giving

$$\sigma_{rr} = 2(\lambda_2 + G_T)A(t) - 2G_T B(t) / r^2 + \lambda_1 C(t),$$

$$\sigma_{\theta\theta} = 2(\lambda_2 + G_T)A(t) + 2G_T B(t) / r^2 + \lambda_1 C(t),$$

$$\sigma_{zz} = 2\lambda_1 A(t) + (\lambda_1 + 2G_1)C(t), \tag{2.62}$$

$$\sigma_{r\theta} = \sigma_{\theta z} = \sigma_{zr} = 0$$

Applying the boundary conditions (2.57)–(2.59), the functions $A(t)$, $B(t)$, and $C(t)$ can be given by

$$A(t) = -\frac{[\pi b^2(\lambda_1 + 2G_1)P_1 + \lambda_1 P_0]}{2\pi(b^2 - a^2)[(\lambda_2 + G_T)(\lambda_1 + 2G_1) - \lambda_1^2]},$$

$$B(t) = -\frac{a^2 b^2 P_1}{2G_T(b^2 - a^2)}, \tag{2.63}$$

$$C(t) = \frac{[\pi b^2 \lambda_1 P_1 + (\lambda_2 + G_T)P_0]}{\pi(b^2 - a^2)[(\lambda_2 + G_T)(\lambda_1 + 2G_1) - \lambda_1^2]}$$

Substituting the solution of $A(t)$, $B(t)$, and $C(t)$ into Equation (2.60), we obtain the expressions of the displacement in terms of the loading P_0 and P_1 as

$$u_r = -\frac{[\pi b^2(\lambda_1 + 2G_1)P_1 + \lambda_1 P_0]r}{2\pi(b^2 - a^2)[(\lambda_2 + G_T)(\lambda_1 + 2G_1) - \lambda_1^2]} - \frac{a^2 b^2 P_1}{2G_T(b^2 - a^2)r},$$

$$u_z = \frac{[\pi b^2 \lambda_1 P_1 + (\lambda_2 + G_T)P_0]z}{\pi(b^2 - a^2)[(\lambda_2 + G_T)(\lambda_1 + 2G_1) - \lambda_1^2]}$$

(2.64)

The strains and stresses can be easily found by substituting Equation (2.63) into Equations (2.61) and (2.62), respectively.

2.4.2 Extension to Poroelastic Bone with Fluid

Papathanasopoulou et al. [29] extended the theory of internal bone remodeling described in Section 2.4.1 to the case of a porous elastic deformable solid in the pores of which a viscous compressible fluid flows. They considered a hollow circular cylinder of poroelastic bone subjected to a quasistatic axial load −$P(t)$ and an internal radial pressure $p(t)$ (see Figure 2.2). The boundary conditions at the inner and outer surfaces of the cylinder are

$$\sigma_{rr} + \sigma = -P(t), \quad \sigma_{r\theta} = \sigma_{rz} = 0 \quad \text{at the inner surface } r = a \qquad (2.65)$$

$$\sigma_{rr} + \sigma = \sigma_{r\theta} = \sigma_{rz} = 0 \quad \text{at the outer surface } r = b \qquad (2.66)$$

The boundary condition at a transverse cross section S of the hollow cylinder is written as

$$\int_S (\sigma_{zz} + \sigma)ds = -P(t) \qquad (2.67)$$

Further, making use of Equations (2.6) and (2.51), the constitutive relations describing the isotropic poroelastic adaptive bone can be written as

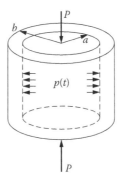

FIGURE 2.2
The poroelastic hollow cylinder and its loading conditions.

$$\sigma_{ij} + \delta_{ij}\sigma = C_{ijkm}(e)\left[\xi\varepsilon_{km} + (1-\xi)\delta_{km}\frac{Q+R}{2N+3A+3Q}\varepsilon\right] \tag{2.68}$$

in which the bulk volume of the poroelastic medium has been assumed to remain constant throughout the remodeling process, which is denoted by the sum of coefficients ξ and $1-\xi$ being equal to unity. Papathanasopoulou et al. [29] further assumed the remodeling rate equation to be in the form of

$$\dot{e} = A_1(e) + A_2(e)\left(E + \frac{Q+R}{2N+3A+3Q}\varepsilon\right) \tag{2.69}$$

where $A_1(e)$ and $A_2(e)$ are material constants dependent upon the change in the volume fraction e.

The system of Equations (2.13), (2.40), (2.46), (2.47), (2.65)–(2.67), (2.68), and (2.69), together with the proper initial conditions, constitute a well-posed initial boundary value problem. Papathanasopoulou et al. proposed the following solution, which satisfies the equations mentioned:

$$u_r = u_r(r,t), \qquad u_\theta = 0, \qquad u_z = D_1(t)z,$$
$$U_r = U_r(r,t), \qquad U_\theta = 0, \qquad U_z = D_2(t)z \tag{2.70}$$

Then, the constitutive equations (2.46) become

$$\sigma_{rr} = (2N+A)\frac{\partial u_r}{\partial r} + A\left(\frac{u_r}{r} + \frac{\partial u_z}{\partial z}\right) + Q\left(\frac{\partial U_r}{\partial r} + \frac{U_r}{r} + \frac{\partial U_z}{\partial z}\right),$$

$$\sigma_{\theta\theta} = (2N+A)\frac{u_r}{r} + A\left(\frac{\partial u_r}{\partial r} + \frac{\partial u_z}{\partial z}\right) + Q\left(\frac{\partial U_r}{\partial r} + \frac{U_r}{r} + \frac{\partial U_z}{\partial z}\right),$$

$$\sigma_{zz} = (2N+A)\frac{\partial u_z}{\partial z} + A\left(\frac{\partial u_r}{\partial r} + \frac{u_r}{r}\right) + Q\left(\frac{\partial U_r}{\partial r} + \frac{U_r}{r} + \frac{\partial U_z}{\partial z}\right), \tag{2.71}$$

$$\sigma = Q\left(\frac{\partial u_r}{\partial r} + \frac{u_r}{r} + \frac{\partial u_z}{\partial z}\right) + R\left(\frac{\partial U_r}{\partial r} + \frac{U_r}{r} + \frac{\partial U_z}{\partial z}\right),$$

$$\sigma_{r\theta} = \sigma_{\theta z} = \sigma_{rz} = 0$$

The equilibrium equation (2.46), in this case, is simplified as

$$\frac{\partial(\sigma_{rr}+\sigma)}{\partial r} + \frac{\sigma_{rr}-\sigma_{\theta\theta}}{r} = 0,$$
$$\frac{\partial(\sigma_{zz}+\sigma)}{\partial z} = 0 \tag{2.72}$$

and Darcy's law (2.47) becomes

$$\frac{\partial \sigma}{\partial r} = C \frac{\partial}{\partial t}(U_r - u_r), \quad \frac{\partial \sigma}{\partial z} = C \frac{\partial}{\partial t}(U_z - u_z) \tag{2.73}$$

Substituting Equation (2.71) into Equations (2.72) and (2.73), we have

$$(2N + A + Q)\aleph u_r + (Q + R)\aleph U_r = 0,$$

$$Q\aleph u_r + R\aleph U_r + C \frac{\partial}{\partial t}(u_r - U_r) = 0 \tag{2.74}$$

and

$$(2N + A + Q)\frac{\partial^2 u_z}{\partial z^2} + (Q + R)\frac{\partial^2 U_z}{\partial z^2} = 0,$$

$$Q\frac{\partial^2 u_z}{\partial z^2} + R\frac{\partial^2 U_z}{\partial z^2} + C \frac{\partial}{\partial t}(u_z - U_z) = 0 \tag{2.75}$$

where

$$\aleph = \frac{\partial^2}{\partial r^2} + \frac{1}{r}\frac{\partial}{\partial r} - \frac{1}{r^2} \tag{2.76}$$

Substituting the assumed solution (2.70) into Equation (2.75), an identity and a relationship are obtained as

$$Cz \frac{\partial}{\partial t}(D_1(t) - D_2(t)) = 0 \tag{2.77}$$

which is equivalent to

$$D_1(t) = D_2(t) + \Theta \tag{2.78}$$

Making use of Equations (2.70), (2.71), and (2.78), the boundary conditions (2.65)–(2.67) become

$$(2N + A + Q)\frac{\partial u_r}{\partial r} + (A + Q)\left(\frac{u_r}{r} - D_1(t)\right)$$

$$+ (Q + R)\left(\frac{\partial U_r}{\partial r} + \frac{U_r}{r} - D_1(t) - \Theta\right) = \begin{cases} -p(t) & \text{at } r = a \\ 0 & \text{at } r = b \end{cases} \tag{2.79}$$

and

$$2\pi(A+Q)(bu_r(b,t)-au_r(a,t))-\pi(b^2-a^2)(2N+A+Q)D_1(t)$$
$$+2\pi(A+Q)(bU_r(b,t)-aU_r(a,t))-\pi(b^2-a^2)(R+Q)(D_1(t)-\Theta)=-P(t) \tag{2.80}$$

at any cross section S.

In handling this problem, having nonhomogeneous boundary conditions, Papathanasopoulou et al. employed the following approach. They assumed that a function $w = w(r,t)$ exists such that

$$\aleph w(r,t)=0 \Rightarrow w(r,t)=A_1(t)r+A_2(t)/r \tag{2.81}$$

which satisfies the conditions

$$(2N+A+Q)\frac{\partial w}{\partial r}+(A+Q)\left(\frac{w}{r}+D_1(t)\right)$$
$$+(Q+R)\left(\frac{\partial w}{\partial r}+\frac{w}{r}+D_1(t)+\Theta\right)=\begin{cases} p(t) & \text{at } r=a \\ 0 & \text{at } r=b \end{cases} \tag{2.82}$$

and

$$2\pi(A+2Q+R)(bw(b,t)-aw(a,t))+\pi(b^2-a^2)(2N+A+2Q+R)D_1(t)$$
$$+\pi(b^2-a^2)(R+Q)\Theta=P(t) \tag{2.83}$$

or, equivalently,

$$2(N+A+2Q+R)A_1(t)-\frac{2N}{a^2}A_2(t)+(A+2Q+R)D_1(t)=p(t)-(R+Q)\Theta \tag{2.84}$$

$$2(N+A+2Q+R)A_1(t)-\frac{2N}{b^2}A_2(t)+(A+2Q+R)D_1(t)=-(R+Q)\Theta \tag{2.85}$$

$$2(A+2Q+R)A_1(t)+(2N+A+2Q+R)D_1(t)=\frac{P(t)}{\pi(b^2-a^2)}-(R+Q)\Theta \tag{2.86}$$

Then, the functions

$$\tilde{u}_r(r,t)=u_r(r,t)+w(r,t), \quad \tilde{U}_r(r,t)=U_r(r,t)+w(r,t) \tag{2.87}$$

satisfy the system of equations

$$(2N + A + Q)\aleph \tilde{u}_r + (Q + R)\aleph \tilde{U}_r = 0,$$

$$Q\aleph \tilde{u}_r + R\aleph \tilde{U}_r + C\frac{\partial}{\partial t}(\tilde{u}_r - \tilde{U}_r) = 0 \qquad (2.88)$$

with homogeneous boundary conditions at $r = a$ and $r = b$:

$$(2N + A + Q)\frac{\partial \tilde{u}_r}{\partial r} + (A + Q)\frac{\tilde{u}_r}{r} + (Q + R)\left(\frac{\partial \tilde{U}_r}{\partial r} + \frac{\tilde{U}_r}{r}\right) = 0 \qquad (2.89)$$

and at S:

$$(A + Q)(b\tilde{u}_r(b,t) - a\tilde{u}_r(a,t)) + (R + Q)(b\tilde{U}_r(b,t) - a\tilde{U}_r(a,t)) = 0 \qquad (2.90)$$

The system of Equation (2.88) can be rewritten as

$$\mathbf{D}\begin{Bmatrix} \tilde{u}_r \\ \tilde{U}_r \end{Bmatrix} = \begin{bmatrix} Q\aleph + C\partial/\partial t & R\aleph - C\partial/\partial t \\ (2N + A + Q)\aleph & (Q + R)\aleph \end{bmatrix}\begin{Bmatrix} \tilde{u}_r \\ \tilde{U}_r \end{Bmatrix} = \begin{Bmatrix} 0 \\ 0 \end{Bmatrix} \qquad (2.91)$$

Introducing a function h such that $\det(\mathbf{D})h = 0$ [30], Papathanasopoulou et al. obtained

$$\left\{(Q + R)\left[Q\aleph + C\frac{\partial}{\partial t}\right] - (2N + A + Q)\left[R\aleph - C\frac{\partial}{\partial t}\right]\right\}\aleph h = 0 \qquad (2.92)$$

Assuming that $h = h_1 + h_2$, Equation (2.92) leads to

$$\aleph h_1 = 0 \quad \text{and} \quad \aleph h_2 + k\partial h_2/\partial t = 0 \qquad (2.93)$$

where

$$k = \frac{C(2N + 2Q + R + A)}{Q^2 - 2NR - AR} \qquad (2.94)$$

The solution of Equation $(2.93)_1$ is obviously

$$h_1(r,t) = A_3(t)r + \frac{A_4(t)}{r} \qquad (2.95)$$

Using the method of variable separation, the function h_2 can be written as

$$h_2(r,t) = R(r)T(t) \tag{2.96}$$

Then, Equation (2.93)$_2$ becomes

$$\frac{1}{R(r)}\frac{d^2R(r)}{dr^2} + \frac{1}{rR(r)}\frac{dR(r)}{dr} - \frac{1}{r^2} = -\frac{k}{T(t)}\frac{dT(t)}{dt} = m^2 \tag{2.97}$$

or, equivalently,

$$k\frac{dT(t)}{dt} + m^2T(t) = 0 \tag{2.98}$$

with the solution

$$T(t) = B_1 e^{-(m^2 t/k)} \tag{2.99}$$

and

$$r^2\frac{d^2R(r)}{dr^2} + r\frac{dR(r)}{dr} - (r^2m^2 + 1)R(r) = 0 \tag{2.100}$$

with the solution

$$R(r) = B_2 I_1(mr) + B_3 K_1(mr) \tag{2.101}$$

where $I_1(mr)$ and $K_1(mr)$ are the modified Bessel functions of the first and second kind, respectively, of order one.

Using the solutions (2.99) and (2.101), Equation (2.96) can be written as

$$h_2(r,t) = (B_6 I_1(mr) + B_7 K_1(mr))e^{-(m^2 t/k)} \tag{2.102}$$

Noting that the functions $\tilde{u}_r$ and $\tilde{U}_r$ can be expressed as

$$\tilde{u}_r = (Q+R)\aleph(h_1 + h_2), \quad \tilde{U}_r = -(2N + A + Q)\aleph(h_1 + h_2) \tag{2.103}$$

and

$$\aleph h_2 = -k\frac{\partial h_2}{\partial t} = (B_6 I_1(mr) + B_7 K_1(mr))m^2 e^{-(m^2 t/k)} \tag{2.104}$$

the expressions for $\tilde{u}_r$ and $\tilde{U}_r$ are given by

$$\tilde{u}_r = (Q+R)(B_6 I_1(mr) + B_7 K_1(mr))m^2 e^{-(m^2 t/k)},$$

$$\tilde{U}_r = -(2N+A+Q)(B_6 I_1(mr) + B_7 K_1(mr))m^2 e^{-(m^2 t/k)} \tag{2.105}$$

Introducing Equation (2.105) into the boundary conditions (2.89) gives

$$B_6 I_1(ma) + B_7 K_1(ma) = 0, \quad B_6 I_1(mb) + B_7 K_1(mb) = 0 \tag{2.106}$$

The system of equations (2.106) has a nontrivial solution only if its determinant is zero:

$$I_1(ma)K_1(mb) - I_1(mb)K_1(ma) = 0 \tag{2.107}$$

Then,

$$B_6 = -B_7 K_1(mb)/I_1(mb) \tag{2.108}$$

Equation (2.107) can be used for determining the constant m. Using Equation (2.87), the radial displacements for the solid and the fluid can be written as

$$u_r = (Q+R)(B_6 I_1(mr) + B_7 K_1(mr))m^2 e^{-(m^2 t/k)} - A_1(t)r - \frac{A_2(t)}{r},$$

$$\tilde{U}_r = -(2N+A+Q)(B_6 I_1(mr) + B_7 K_1(mr))m^2 e^{-(m^2 t/k)} - A_1(t)r - \frac{A_2(t)}{r} \tag{2.109}$$

Making use of Equation (2.109) and the expressions for the axial displacements,

$$u_z(t) = -D_1(t)z, \quad U_z(t) = -D_2(t)z \tag{2.110}$$

The stress components are obtained from Equation (2.71) as

$$\sigma_{rr} = -2N(Q+R)\frac{m^2}{r}(B_6 I_1(mr) + B_7 K_1(mr))e^{-(m^2 t/k)}$$

$$- m^3(Q^2 - 2NR - AR)(B_6 I_0(mr) - B_7 K_0(mr))e^{-(m^2 t/k)} \tag{2.111}$$

$$- 2(N+A+Q)A_1(t) + \frac{2NA_2(t)}{r^2} - (A+Q)D_1(t) - Q\Theta$$

$$\sigma_{zz} = -(2QN - AR + Q^2)(B_6 I_0(mr) - B_7 K_0(mr))m^3 e^{-(m^2 t/k)}$$
$$- 2(A+Q)A_1(t) - (2N+A+Q)D_1(t) - Q\Theta \tag{2.112}$$

$$\sigma = (Q^2 - 2RN - AR)(B_6 I_0(mr) - B_7 K_0(mr))m^3 e^{-(m^2 t/k)}$$
$$- 2(R+Q)A_1(t) - (R+Q)D_1(t) - R\Theta \tag{2.113}$$

where $I_0(mr)$ and $K_0(mr)$ are the modified Bessel functions of the first and second kind of zero order, respectively.

The solution of the system of equations (2.84)–(2.86) leads to the expressions for the unknowns, $A_1(t)$, $A_2(t)$, and $D_1(t)$ as

$$A_1 = \frac{(A+2Q+R)P + a^2\pi(A+2N+2Q+R)p + 2N\pi(b^2 - a^2)(Q+R)\Theta}{2(a^2 - b^2)N\pi(3A+2N+6Q+3R)},$$

$$A_2 = \frac{-a^2 b^2 p}{2N(b^2 - a^2)},$$

$$D_1 = \frac{-(A+2Q+R+N)P - a^2\pi(A+2Q+R)p + N\pi(b^2 - a^2)(Q+R)\Theta}{(a^2 - b^2)(3A+2N+6Q+3R)} \tag{2.114}$$

Taking [29]

$$\Theta = \frac{3A+2N+6Q+3R}{Q+R} \quad \text{and} \quad B_7 = 1 \tag{2.115}$$

and substituting Equation (2.114) into Equations (2.109)–(2.113), the displacements and stresses in the hollow cylinder can be obtained.

The formulations discussed in this chapter provide an initial insight into bone remodeling theory. Extensions to multifield bone remodeling analysis are discussed in the next two chapters.

References

1. Wolff J. *Das gesetz der transformation der knochen.* Berlin: A. Hirschwald (1892).
2. Bauman W. A., Spungen A. M., Wang J., Pierson R. N., Schwartz E. Continuous loss of bone during chronic immobilization: A monozygotic twin study. *Osteoporosis International* 10 (2): 123–127 (1999).
3. Zerwekh J. E., Ruml L. A., Gottschalk F., Pak C. Y. C. The effects of twelve weeks of bed rest on bone histology, biochemical markers of bone turnover, and

calcium homeostasis in eleven normal subjects. *Journal of Bone Mineral Research* 13 (10): 1594–1601 (1998).

4. Suominen H. Bone-mineral density and long-term exercise—An overview of cross-sectional athlete studies. *Sports Medicine* 16 (5): 316–330 (1993).

5. Glucksmann A. Studies on bone mechanics in vitro. *Anatomical Record* 72: 97–115 (1938).

6. Bassett C. A. L., Becker R. O. Generation of electric potentials by bone in response to mechanical stress. *Science* 137 (3535): 1063–1064 (1962).

7. Shamos M. H., Shamos M. I., Lavine L. S. Piezoelectric effect in bone. *Nature* 197 (486): 81 (1963).

8. Justus R., Luft J. H. A mechanochemical hypothesis for bone remodeling induced by mechanical stress. *Calcified Tissue Research* 5 (3): 222–235 (1970).

9. Cowin S. C. Adaptive elasticity. *Bulletin of the Australian Mathematical Society* 26 (1): 57–80 (1982).

10. Qin Q. H., Ye J. Q. Thermoelectroelastic solutions for internal bone remodeling under axial and transverse loads. *International Journal of Solids and Structures* 41 (9–10): 2447–2460 (2004).

11. Cowin S. C. The mechanical and stress adaptive properties of bone. *Annals of Biomedical Engineering* 11 (3–4): 263–295 (1983).

12. Frost H. M. Skeletal structural adaptations to mechanical usage (SATMU). 1. Redefining Wolff law—The bone modeling problem. *Anatomical Record* 226 (4): 403–413 (1990).

13. Frost H. M. Skeletal structural adaptations to mechanical usage (SATMU). 2. Redefining Wolff law—The remodeling problem. *Anatomical Record* 226 (4): 414–422 (1990).

14. Cowin S. C., Hegedus D. H. Bone remodeling I: Theory of adaptive elasticity. *Journal of Elasticity* 6 (3): 313–326 (1976).

15. Hegedus D. H., Cowin S. C. Bone remodeling II: small strain adaptive elasticity. *Journal of Elasticity* 6 (4): 337–352 (1976).

16. Cowin S. C., Nachlinger R. R. Bone remodeling III: Uniqueness and stability in adaptive elasticity theory. *Journal of Elasticity* 8 (3): 285–295 (1978).

17. Frost H. M. Dynamics of bone remodeling. In *Bone biodynamics*, ed. Frost H. M. Boston: Little and Brown, p. 316 (1964).

18. Currey J. D. Differences in the blood-supply of bone of different histological types. *Quarterly Journal of Microscopical Science* 101 (3): 351–370 (1960).

19. Woo S. L. Y., Kuei S. C., Amiel D., Gomez M. A., Hayes W. C., White F. C., Akeson W. H. The effect of prolonged physical training on the properties of long-bone—A study of Wolff's law. *Journal of Bone and Joint Surgery,* American vol. 63 (5): 780–787 (1981).

20. Uhthoff H. K., Jaworski Z. F. G. Bone loss in response to long-term immobilization. *Journal of Bone and Joint Surgery,* British vol. 60 (3): 420–429 (1978).

21. Jaworski Z. F. G., Liskovakiar M., Uhthoff H. K. Effect of long-term immobilization on the pattern of bone loss in older dogs. *Journal of Bone and Joint Surgery,* British vol. 62 (1): 104–110 (1980).

22. Goodman M. A., Cowin S. C. Continuum theory for granular materials. *Archive for Rational Mechanics and Analysis* 44 (4): 249–266 (1972).

23. Cowin S. C., Van Buskirk W. C. Surface bone remodeling induced by a medullary pin. *Journal of Biomechanics* 12 (4): 269–276 (1979).

24. Papathanasopoulou V. A., Fotiadis D. I., Massalas C. V. A theoretical analysis of surface remodeling in long bones. *International Journal of Engineering Science* 42 (3–4): 395–409 (2004).
25. Biot M. A. Theory of elasticity and consolidation for a porous anisotropic solid. *Journal of Applied Physics* 26 (2): 182–185 (1955).
26. Johnson M. W., Chakkalakal D. A., Harper R. A., Katz J. L., Rouhana S. W. Fluid flow in bone in vitro. *Journal of Biomechanics* 15 (11): 881–885 (1982).
27. Cowin S. C., Vanbuskirk W. C. Internal bone remodeling induced by a medullary pin. *Journal of Biomechanics* 11 (5): 269–275 (1978).
28. Tsili M. C. Theoretical solutions for internal bone remodeling of diaphyseal shafts using adaptive elasticity theory. *Journal of Biomechanics* 33 (2): 235–239 (2000).
29. Papathanasopoulou V. A., Fotiadis D. I., Foutsitzi G., Massalas C. V. A poroelastic bone model for internal remodeling. *International Journal of Engineering Science* 40 (5): 511–530 (2002).
30. Fotiadis D. I., Foutsitzi G., Massalas C. V. Wave propagation modeling in human long bones. *Acta Mechanica* 137 (1–2): 65–81 (1999).

3

Multifield Internal Bone Remodeling

3.1 Introduction

In the previous two chapters some fundamental concepts and basic formulations for bone remodeling processes have been presented. We now present applications of these formulations to multifield internal bone remodeling of inhomogeneous long cylindrical bone. Bone remodeling processes are the mechanisms by which bone adapts its histological structure to changes in long-term loading. As explained in Chapter 2, there are two kinds of bone remodeling: internal and surface [1]. This chapter describes multifield internal bone remodeling; then, Chapter 4 gives the theory and solution of multifield surface bone remodeling.

The capacity of bone remodeling has been investigated by many authors [2–14]. Active research in the area of bone tissue such as living bone and collagen has also shown these materials to be piezoelectric [15,16], and the piezoelectric properties of bone play an important role in the development and growth of remodeling of skeletons. Applications of piezoelectric theory to bone remodeling have been the subject of fruitful scientific attention by many distinguished researchers (e.g., references 17–19 and others). In particular, Gjelsvik [20] presented a physical description of the remodeling of bone tissue in terms of a very simplified form of linear theory of piezoelectricity. Williams and Breger [21] explored the applicability of stress gradient theory for explaining the experimental data for a cantilever bone beam subjected to constant end load, showing that the approximate gradient theory was in good agreement with the experimental data. Guzelsu [22] presented a piezoelectric model for analyzing a cantilever dry bone beam subjected to a vertical end load.

Johnson, Williams, and Gross [23] further addressed the problem of a dry bone beam by presenting some theoretical expressions for the piezoelectric response to cantilever bending of the beam. Demiray [24] provided a theoretical description of electromechanical remodeling models of bones. Aschero et al. [25] investigated the converse piezoelectric effect of fresh bone using a highly sensitive dilatometer. They further investigated the piezoelectric properties of bone and presented a set of repeated measurements of coefficient d_{23} in 25 cow bone samples [26]. Fotiadis, Foutsitzi, and Massalas [27] studied wave propagation in a long cortical piezoelectric bone with arbitrary cross section. El-Naggar and Abd-Alla [28] and Ahmed and Abd-Alla [29] further obtained an analytical solution for wave propagation in long cylindrical bones with and without a cavity. Silva et al. [30] explored the physicochemical, dielectric, and piezoelectric properties of anionic collagen and collagen-hydroxyapatite composites. Recently, Qin and Ye [19] and Qin, Qu, and Ye [9] presented a thermoelectroelastic solution for internal and surface bone remodeling, respectively. Later, Qu and Qin [11] and Qu, Qin, and Kang [14] extended the results of Qin et al. [9] and Qin and Ye [19] to include magnetic effects. Accounts of most of the developments in this area can also be found in references 31 and 32. In this chapter, however, we restrict our discussion to the findings presented in references 11, 19, and 31.

3.2 Linear Theory of Thermoelectroelastic Bone

Consider a hollow circular cylinder composed of linearly thermopiezoelectric bone material subjected to axisymmetric loading. The axial, circumferential, and normal to the middle surface coordinate length parameters are denoted by z, θ, and r, respectively. In the case of thermoelectroelastic bone, the constitutive equations (2.11) now become [33]

$$\sigma_{rr} = c_{11}\varepsilon_{rr} + c_{12}\varepsilon_{\theta\theta} + c_{13}\varepsilon_{zz} - e_{31}E_z - \lambda_1 T$$

$$\sigma_{\theta\theta} = c_{12}\varepsilon_{rr} + c_{11}\varepsilon_{\theta\theta} + c_{13}\varepsilon_{zz} - e_{31}E_z - \lambda_1 T$$

$$\sigma_{zz} = c_{13}\varepsilon_{rr} + c_{13}\varepsilon_{\theta\theta} + c_{33}\varepsilon_{zz} - e_{33}E_z - \lambda_3 T \qquad (3.1)$$

$$\sigma_{zr} = c_{44}\varepsilon_{zr} - e_{15}E_r, \qquad D_r = e_{15}\varepsilon_{zr} + \kappa_1 E_r$$

$$D_z = d_1(\varepsilon_{rr} + \varepsilon_{\theta\theta}) + e_{33}\varepsilon_{zz} + \kappa_3 E_z - \chi_3 T$$

$$h_r = k_r H_r, \qquad h_z = k_z H_z$$

where
    σ_{ij} and ε_{ij} again represent components of stress and strain
    D_i and E_i denote components of electric displacement and electric field
        intensity, respectively
    c_{ij} is elastic stiffness
    e_{ij} is a piezoelectric constant
    κ_i is the dielectric permittivity
    T denotes temperature change
    χ_3 is a pyroelectric constant
    λ_i is a stress-temperature coefficient

h_i is heat flow
H_i is heat intensity
k_i is heat conduction coefficient

The associated strain–displacement relations are again defined by Equation (2.13), and the related electric fields and heat intensities are respectively related to electric potential φ and temperature change T as

$$E_r = -\varphi_{,r}, \ E_z = -\varphi_{,z}, \ H_r = -T_{,r}, \ H_z = -T_{,z} \tag{3.2}$$

For quasistationary behavior, in the absence of a heat source, free electric charge, and body forces, the set of equations for the thermopiezoelectric theory of bone is completed by adding the following equations of equilibrium for heat flow, stress, and electric displacements to Equations (3.1) and (3.2):

$$\frac{\partial \sigma_{rr}}{\partial r} + \frac{\partial \sigma_{zr}}{\partial z} + \frac{\sigma_{rr} - \sigma_{\theta\theta}}{r} = 0, \qquad \frac{\partial \sigma_{zr}}{\partial r} + \frac{\partial \sigma_{zz}}{\partial z} + \frac{\sigma_{zr}}{r} = 0,$$

$$\frac{\partial D_r}{\partial r} + \frac{\partial D_z}{\partial z} + \frac{D_r}{r} = 0, \qquad \frac{\partial h_r}{\partial r} + \frac{\partial h_z}{\partial z} + \frac{h_r}{r} = 0, \tag{3.3}$$

where Equation (2.14) is rewritten here as the first two terms of Equation (3.3) for readers' convenience.

The bone remodeling rate equation (2.7) is modified by adding some additional terms related to electric fields as

$$\dot{e} = A(e) + A_r^E(e)E_r + A_z^E(e)E_z + A_{rr}^\varepsilon(\varepsilon_{rr} + \varepsilon_{\theta\theta}) + A_{zz}^\varepsilon \varepsilon_{zz} + A_{zr}^\varepsilon \varepsilon_{rz} \tag{3.4}$$

where $A_i^E(e)$ and $A_{ij}^\varepsilon(e)$ are material constants dependent upon the volume function e. Equations (3.1)–(3.4) form the basic set of equations for the adaptive theory of internal piezoelectric bone remodeling.

3.3 Analytical Solution of a Homogeneous Hollow Circular Cylindrical Bone

We now consider a hollow circular cylinder of bone subjected to an external temperature change T_0, a quasistatic axial pressure load P, an external pressure p, and an electric load φ_a(or/and φ_b) as shown in Figure 3.1. The boundary conditions are

$$T = 0, \qquad \sigma_{rr} = \sigma_{r\theta} = \sigma_{rz} = 0, \qquad \varphi = \varphi_a \qquad \text{at } r = a$$

$$T = T_0, \qquad \sigma_{rr} = -p, \quad \sigma_{r\theta} = \sigma_{rz} = 0, \quad \varphi = \varphi_b, \quad \text{at } r = b \tag{3.5}$$

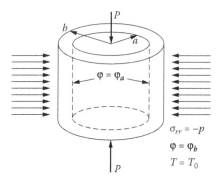

FIGURE 3.1
Geometry and loading of the hollow cylinder.

and

$$\int_S \sigma_{zz} dS = -P \tag{3.6}$$

where a and b denote, respectively, the inner and outer radii of the bone, and S is the cross-sectional area. For a long bone, it is assumed that all displacements, temperature, and electrical potential except the axial displacement u_z are independent of the z coordinate and that u_z may have linear dependence on z. Using Equations (3.1) and (3.2), differential equations (3.3) can be rewritten as

$$\left(\frac{\partial^2}{\partial r^2} + \frac{1}{r} \frac{\partial}{\partial r} \right) T = 0, \quad c_{11} \left(\frac{\partial^2}{\partial r^2} + \frac{1}{r} \frac{\partial}{\partial r} - \frac{1}{r^2} \right) u_r = \lambda_1 \frac{\partial T}{\partial r} \tag{3.7}$$

$$\left(\frac{\partial^2}{\partial r^2} + \frac{1}{r} \frac{\partial}{\partial r} \right) (c_{44} u_z + e_{15} \varphi) = 0 \tag{3.8}$$

$$\left(\frac{\partial^2}{\partial r^2} + \frac{1}{r} \frac{\partial}{\partial r} \right) (e_{15} u_z - \kappa_1 \varphi) = 0 \tag{3.9}$$

The solution to the heat conduction Equation $(3.7)_1$ satisfying boundary conditions (3.5) can be written as

$$T = \frac{\ln(r / a)}{\ln(b / a)} T_0 \tag{3.10}$$

It is easy to prove that Equations (3.7)–(3.9) will be satisfied if we assume

$$u_r = A(t)r + \frac{B(t)}{r} + \frac{\varpi r[\ln(r / a) - 1]}{c_{11}} \tag{3.11}$$

$$u_z = zC(t) + D(t)\ln(r/a), \quad \varphi = F(t)\ln(r/a) + \varphi_a \tag{3.12}$$

where $A, B, C, D,$ and F are unknown variables to be determined by introducing boundary conditions, and $\varpi = \dfrac{\lambda_1 T_0}{2\ln(b/a)}$. Substituting Equations (3.11) and (3.12) into Equation (3.2) and later into Equation (3.1), we obtain

$$\sigma_{rr} = A(t)(c_{11} + c_{12}) - \frac{B(t)}{r^2}(c_{11} - c_{12}) + c_{13}C(t) + \varpi\left[\frac{c_{12}}{c_{11}}\left(\ln\frac{r}{a} - 1\right) - \ln\frac{r}{a}\right] \tag{3.13}$$

$$\sigma_{\theta\theta} = A(t)(c_{11} + c_{12}) + \frac{B(t)}{r^2}(c_{11} - c_{12}) + c_{13}C(t) + \varpi\left[\frac{c_{12}}{c_{11}}\ln\frac{r}{a} - \ln\frac{r}{a} - 1\right] \tag{3.14}$$

$$\sigma_{zz} = 2A(t)c_{13} + c_{33}C(t) + \varpi\frac{c_{13}}{c_{11}}[2\ln(r/a) - 1] - \lambda_3 T_0\frac{\ln(r/a)}{\ln(b/a)} \tag{3.15}$$

$$\sigma_{zr} = \frac{1}{r}[c_{44}D(t) + e_{15}F(t)], \quad D_r = \frac{1}{r}[e_{15}D(t) - \kappa_1 F(t)] \tag{3.16}$$

$$D_z = 2A(t)e_{31} + C(t)e_{33} + \varpi\frac{e_{31}}{c_{11}}[2\ln(r/a) - 1] - \chi_3 T_0\frac{\ln(r/a)}{\ln(b/a)} \tag{3.17}$$

The boundary conditions (3.5) and (3.6) of stresses and electric potential require that

$$c_{44}D(t) + d_4 F(t) = 0, \quad \varphi_b = F(t)\ln(b/a) + \varphi_a \tag{3.18}$$

$$A(t)(c_{11} + c_{12}) - \frac{B(t)}{a^2}(c_{11} - c_{12}) + c_{13}C(t) - \frac{c_{12}}{c_{11}}\varpi = 0 \tag{3.19}$$

$$A(t)(c_{11} + c_{12}) - \frac{B(t)}{b^2}(c_{11} - c_{12}) + c_{13}C(t) + \varpi\left[\frac{c_{12}}{c_{11}}\left(\ln\frac{b}{a} - 1\right) - \ln\frac{b}{a}\right] = -p \tag{3.20}$$

$$\pi(b^2 - a^2)[2A(t)c_{13} + C(t)c_{33} - F_1^* T_0] + F_2^* T_0 = -P \tag{3.21}$$

where

$$F_1^* = \frac{1}{\ln(b/a)}\left(\frac{c_{13}\lambda_1}{c_{11}} - \frac{\lambda_3}{2}\right), \quad F_2^* = \pi b^2\left(\frac{c_{13}}{c_{11}}\lambda_1 - \lambda_3\right) \tag{3.22}$$

The unknown functions $A(t), B(t), C(t), D(t),$ and $F(t)$ are readily found from Equations (3.18)–(3.21) as

$$A(t) = \frac{1}{F_3^*}\left(c_{33}\beta_1^*[\beta_2^* T_0 + p(t)] + \varpi\frac{c_{33}c_{12}}{c_{11}} + \frac{F_2^* T_0 + P(t)}{\pi(b^2 - a^2)}c_{13} - F_1^* T_0 c_{13}\right) \tag{3.23}$$

$$B(t) = \frac{a^2 \beta_1^* [\beta_2^* T_0 + p(t)]}{c_{11} - c_{12}} \tag{3.24}$$

$$C(t) = \frac{1}{F_3^*} \left(\left[F_1^* T_0 - \frac{F_2^* T_0 + P(t)}{\pi(b^2 - a^2)} \right] (c_{11} + c_{12}) - 2c_{13}\beta_1^* [\beta_2^* T_0 + p(t)] - \frac{2c_{13}c_{12}\varpi}{c_{11}} \right) \tag{3.25}$$

$$D(t) = -\frac{e_{15}(\varphi_b - \varphi_a)}{c_{44} \ln(b/a)} \tag{3.26}$$

$$F(t) = \frac{\varphi_b - \varphi_a}{\ln(b/a)} \tag{3.27}$$

where

$$F_3^* = c_{33}(c_{11} + c_{12}) - 2c_{13}^2, \quad \beta_1^* = \frac{b^2}{(a^2 - b^2)}, \quad \beta_2^* = \frac{\lambda_1}{2}\left(\frac{c_{12}}{c_{11}} - 1\right) \tag{3.28}$$

Using Equations (3.23)–(3.27), the displacements u_r, u_z, and electrical potential φ are given by

$$u_r = \frac{r}{F_3^*} \left(c_{33}\beta_1^*[\beta_2^* T_0 + p(t)] + \varpi \frac{c_{33}c_{12}}{c_{11}} + \frac{F_2^* T_0 + P(t)}{\pi(b^2 - a^2)}c_{13} - F_1^* T_0 c_{13} \right)$$
$$+ \frac{a^2 \beta_1^*[\beta_2^* T_0 + p(t)]}{r(c_{11} - c_{12})} + \frac{\varpi r[\ln(r/a) - 1]}{c_{11}} \tag{3.29}$$

$$u_z = \frac{z}{F_3^*} \left(\left[F_1^* T_0 - \frac{F_2^* T_0 + P(t)}{\pi(b^2 - a^2)} \right](c_{11} + c_{12}) - 2c_{13}\beta_1^*[\beta_2^* T_0 + p(t)] \right.$$
$$\left. - \frac{2c_{13}c_{12}\varpi}{c_{11}} \right) - \frac{e_{15}(\varphi_b - \varphi_a)\ln(r/a)}{c_{44}\ln(b/a)} \tag{3.30}$$

$$\varphi = \frac{\ln(r/a)}{\ln(b/a)}(\varphi_b - \varphi_a) + \varphi_a \tag{3.31}$$

The strains and electric field intensity appearing in Equation (3.4) can be found by substituting Equations (3.29)–(3.31) into Equation (3.2). They are, respectively,

$$\varepsilon_{rr} = \frac{1}{F_3^*} \left(c_{33}\beta_1^*[\beta_2^* T_0 + p(t)] + \varpi \frac{c_{33}c_{12}}{c_{11}} + \frac{F_2^* T_0 + P(t)}{\pi(b^2 - a^2)}c_{13} - F_1^* T_0 c_{13} \right)$$
$$- \frac{a^2 \beta_1^*[\beta_2^* T_0 + p(t)]}{r^2(c_{11} - c_{12})} + \frac{\varpi \ln(r/a)}{c_{11}} \tag{3.32}$$

$$\varepsilon_{\theta\theta} = \frac{1}{F_3^*}\left(c_{33}\beta_1^*[\beta_2^*T_0 + p(t)] + \varpi\frac{c_{33}c_{12}}{c_{11}} + \frac{F_2^*T_0 + P(t)}{\pi(b^2 - a^2)}c_{13} - F_1^*T_0c_{13} \right)$$

$$+ \frac{a^2\beta_1^*[\beta_2^*T_0 + p(t)]}{r^2(c_{11} - c_{12})} + \frac{\varpi[\ln(r/a) - 1]}{c_{11}} \tag{3.33}$$

$$\varepsilon_{zz} = \frac{1}{F_3^*}\left(\left[F_1^*T_0 - \frac{F_2^*T_0 + P(t)}{\pi(b^2 - a^2)} \right](c_{11} + c_{12}) - 2c_{13}\beta_1^*[\beta_2^*T_0 + p(t)] \right.$$

$$\left. - \frac{2c_{13}c_{12}\varpi}{c_{11}} \right) \tag{3.34}$$

$$\varepsilon_{rz} = -\frac{e_{15}(\varphi_b - \varphi_a)}{rc_{44}\ln(b/a)} \tag{3.35}$$

$$E_r = -\frac{(\varphi_b - \varphi_a)}{r\ln(b/a)} \tag{3.36}$$

Then, substituting the solutions (3.32)–(3.36) into Equation (3.4) yields

$$\dot{e} = A^*(e) + \frac{2A_{rr}^\varepsilon}{F_3^*}\left(c_{33}\beta_1^*[\beta_2^*T_0 + p(t)] + \varpi\frac{c_{33}c_{12}}{c_{11}} + \frac{F_2^*T_0 + P(t)}{\pi(b^2 - a^2)}c_{13} - F_1^*T_0c_{13} \right)$$

$$+ \frac{A_{rr}^\varepsilon\varpi[2\ln(r/a) - 1]}{c_{11}} + \frac{A_{zz}^\varepsilon}{F_3^*}\left(\left[F_1^*T_0 - \frac{F_2^*T_0 + P(t)}{\pi(b^2 - a^2)} \right](c_{11} + c_{12}) \right. \tag{3.37}$$

$$\left. - 2c_{13}\beta_1^*[\beta_2^*T_0 + p(t)] - \frac{2c_{13}c_{12}\varpi}{c_{11}} \right) - \frac{\varphi_b - \varphi_a}{r\ln(b/a)}\left(A_r^E + \frac{e_{15}}{c_{44}}A_{zr}^\varepsilon \right)$$

Since we do not know the exact expressions of the material functions $A^*(e)$, $A_i^E(e)$, $A_{ij}^\varepsilon(e)$, c_{ij}, d_j, λ_j, κ_j, and χ_3, the following approximate forms of them, as proposed by Cowin and Van Buskirk [4] for small values of e, are used here:

$$A^*(e) = C_0 + C_1e + C_2e^2, \quad A_i^E(e) = A_i^{E0} + eA_i^{E1}, \quad A_{ij}^\varepsilon(e) = A_{ij}^{\varepsilon0} + eA_{ij}^{\varepsilon1} \tag{3.38}$$

and

$$c_{ij}(e) = c_{ij}^0 + \frac{e}{\xi_0}(c_{ij}^1 - c_{ij}^0), \quad e_{ij}(e) = e_{ij}^0 + \frac{e}{\xi_0}(e_{ij}^1 - e_{ij}^0),$$

$$\lambda_i(e) = \lambda_i^0 + \frac{e}{\xi_0}(\lambda_i^1 - \lambda_i^0), \quad \kappa_i(e) = \kappa_i^0 + \frac{e}{\xi_0}(\kappa_i^1 - \kappa_i^0), \tag{3.39}$$

$$\chi_3(e) = \chi_3^0 + \frac{e}{\xi_0}(\chi_3^1 - \chi_3^0)$$

where $C_0, C_1, C_2, A_i^{E0}, A_i^{E1}, A_{ij}^{\varepsilon 0}, A_{ij}^{\varepsilon 1}, c_{ij}^0, c_{ij}^1, e_{ij}^0, e_{ij}^1, \lambda_i^0, \lambda_i^1, \kappa_i^0, \kappa_i^1, \chi_3^0,$ and χ_3^1 are material constants. Using these approximations, the remodeling rate equation (3.37) can be simplified as

$$\dot{e} = \alpha(e^2 - 2\beta e + \gamma) \tag{3.40}$$

by neglecting terms of e^3 and the higher orders of e, where α, β, and γ are constants. The solution to Equation (3.40) is straightforward and has been discussed by Hegedus and Cowin [34]. For the reader's convenience, the solution process is briefly described here. Let e_1 and e_2 denote solutions to $e^2 - 2\beta e + \gamma = 0$ that is,

$$e_{1,2} = \beta \pm (\beta^2 - \gamma)/^{1/2} \tag{3.41}$$

When $\beta^2 < \gamma$, e_1 and e_2 are a pair of complex conjugates and the solution of Equation (3.40) is

$$e(t) = \beta + \sqrt{(\gamma - \beta^2)} \tan\left(\alpha t \sqrt{(\gamma - \beta^2)} + \arctan \frac{\sqrt{(\gamma - \beta^2)}}{\beta - e_0} \right) \tag{3.42}$$

where $e = e_0$ is the initial condition. When $\beta^2 = \gamma$, the solution is

$$e(t) = e_1 - \frac{e_1 - e_0}{1 + \alpha(e_1 - e_0)t} \tag{3.43}$$

Finally, when $\beta^2 > \gamma$, we have

$$e(t) = \frac{e_1(e_0 - e_2) + e_2(e_1 - e_0)\exp(\alpha(e_1 - e_2)t)}{(e_0 - e_2) + (e_1 - e_0)\exp(\alpha(e_1 - e_2)t)} \tag{3.44}$$

Since it has been proved that both solutions (3.42) and (3.43) are physically unlikely [4], we will use solution (3.44) in our numerical analysis.

3.4 Semianalytical Solution for Inhomogeneous Cylindrical Bone Layers

The solution obtained in the previous section is suitable for analyzing bone cylinders if they are assumed to be homogeneous [4]. It can be useful if explicit expressions and a simple analysis are required. In fact, however, all bone materials exhibit inhomogeneity. In particular, for a hollow bone cylinder, the volume fraction of bone matrix materials varies from the inner to the outer surface. To solve this problem, a semianalytical model was presented in Qin and Ye [19]. The following discussion is a brief summary of this work.

Considering Equations (3.1)–(3.3) and assuming a constant longitudinal strain, the following first-order differential equations can be obtained [19]:

$$
\frac{\partial}{\partial r}
\begin{Bmatrix} u_r \\ \sigma_r \end{Bmatrix}
=
\begin{bmatrix}
-\dfrac{c_{12}}{c_{11}r} & \dfrac{1}{c_{11}} \\
\dfrac{\psi}{r^2} & \dfrac{(c_{12}/c_{11}-1)}{r}
\end{bmatrix}
\begin{Bmatrix} u_r \\ \sigma_r \end{Bmatrix}
$$

$$
+
\begin{Bmatrix}
-\dfrac{c_{13}}{c_{11}} \\
\dfrac{c_{13}(1-c_{12}/c_{11})}{r}
\end{Bmatrix}
\varepsilon_z
+
\begin{Bmatrix}
\dfrac{\lambda_1}{c_{11}} \\
\dfrac{(c_{12}/c_{11}-1)\lambda_1}{r}
\end{Bmatrix}
T
\tag{3.45}
$$

where $\psi = c_{11} - c_{12}^2/c_{11}$. In this equation, the effect of electrical potential is absent because it is independent of u_r and σ_r. The contribution of electrical field can be calculated separately, as described in the previous section, and then included in the remodeling rate equation.

Assuming that a bone layer is sufficiently thin, we can replace r with its mean value R and let $r = a + s$, where $0 \le s \le h$, and a and h are the inner radius and the thickness of the thin bone layer, respectively. Thus, Equation (3.45) is reduced to

$$
\frac{\partial}{\partial s}
\begin{Bmatrix} u_r \\ \sigma_r \end{Bmatrix}
=
\begin{bmatrix}
-\dfrac{c_{12}}{c_{11}R} & \dfrac{1}{c_{11}} \\
\dfrac{\psi}{R^2} & \dfrac{(c_{12}/c_{11}-1)}{R}
\end{bmatrix}
\begin{Bmatrix} u_r \\ \sigma_r \end{Bmatrix}
$$

$$
+
\begin{Bmatrix}
-\dfrac{c_{13}}{c_{11}} \\
\dfrac{c_{13}(1-c_{12}/c_{11})}{R}
\end{Bmatrix}
\varepsilon_z
+
\begin{Bmatrix}
\dfrac{\lambda_1}{c_{11}} \\
\dfrac{(c_{12}/c_{11}-1)\lambda_1}{R}
\end{Bmatrix}
T
\tag{3.46}
$$

This equation can be written symbolically as

$$
\frac{\partial}{\partial s}\{\mathbf{F}\} = [\mathbf{G}]\{\mathbf{F}\} + \{\mathbf{H}_L\} + \{\mathbf{H}_T\}
\tag{3.47}
$$

where $[\mathbf{G}]$, $\{\mathbf{H}_L\}$, and $\{\mathbf{H}_T\}$ are all constant matrices.

Equation (3.47) can be solved analytically and the solution is [35]

$$\begin{Bmatrix} u_r(s) \\ \sigma_r(s) \end{Bmatrix} = e^{[\mathbf{G}]s} \begin{Bmatrix} u_r(0) \\ \sigma_r(0) \end{Bmatrix} + \int_0^h e^{[\mathbf{G}](s-\tau)} \{\mathbf{H}_L\} d\tau + \int_0^h e^{[\mathbf{G}](s-\tau)} \{\mathbf{H}_T\} d\tau \qquad (3.48)$$

where $u_r(0)$ and $\sigma_r(0)$ are, respectively, displacement and stress at the bottom surface of the layer. Rewrite Equation (3.48) as

$$\{\mathbf{F}(s)\} = [\mathbf{D}(s)]\{\mathbf{F}(0)\} + \{\mathbf{D}_L\} + \{\mathbf{D}_T\} \qquad (3.49)$$

The exponential matrix can be calculated as follows:

$$[\mathbf{D}(s)] = e^{[\mathbf{G}]s} = \alpha_0(s)\mathbf{I} + \alpha_1(s)[\mathbf{G}] \qquad (3.50)$$

where $\alpha_0(s)$ and $\alpha_1(s)$ can be solved from

$$\alpha_0(s) + \alpha_1(s)\beta_1 = e^{\beta_1 s}$$
$$\alpha_0(s) + \alpha_1(s)\beta_2 = e^{\beta_2 s} \qquad (3.51)$$

In Equation (3.51), β_1 and β_2 are two eigenvalues of $[\mathbf{G}]$, which are given by

$$\begin{Bmatrix} \beta_1 \\ \beta_2 \end{Bmatrix} = -\frac{1}{2R} \pm \frac{1}{2R}\sqrt{5 - 4\frac{c_{12}}{c_{11}}} \qquad (3.52)$$

Considering now $s = h$ (i.e., the external surface of the bone layer), we obtain

$$\{\mathbf{F}(h)\} = [\mathbf{D}(h)]\{\mathbf{F}(0)\} + \{\mathbf{D}_L\} + \{\mathbf{D}_T\} \qquad (3.53)$$

The axial stress applied at the end of the bone can be found as

$$\sigma_z = c_{13}\left(1 - \frac{c_{12}}{c_{11}}\right)\frac{u}{R} + \left(c_{33} - \frac{c_{13}^2}{c_{11}}\right)\varepsilon_z + \frac{c_{13}}{c_{11}}\sigma_r + \left(\frac{c_{13}}{c_{11}}\lambda_1 - \lambda_3\right)T \qquad (3.54)$$

The stress problem (3.54) can be solved by introducing the boundary conditions described on the top and bottom surfaces into Equation (3.53) and

$$\iint_S \left\{ c_{13}\left(1-\frac{c_{12}}{c_{11}}\right)\frac{u}{R} + \left(c_{33}-\frac{c_{13}^2}{c_{11}}\right)\varepsilon_z + \frac{c_{13}}{c_{11}}\sigma_r + \left(\frac{c_{13}}{c_{11}}\lambda_1 - \lambda_3\right)T\right\}dS = -P(t) \quad (3.55)$$

For a thick-walled bone section or a section with variable volume fraction in the radial direction, we can divide the bone into a number of sublayers, each of which is sufficiently thin and is assumed to be composed of a homogeneous material. Within a layer we take the mean value of the volume fraction of the layer as the layer's volume fraction. As a consequence, the analysis described previously for a thin and homogeneous bone can be applied here for each sublayer in a straightforward manner. For instance, for the *j*th layer, Equation (3.53) becomes

$$\left\{\mathbf{F}^{(j)}(h_j)\right\} = \left[\mathbf{D}^{(j)}(h_j)\right]\left\{\mathbf{F}^{(j)}(0)\right\}_j + \left\{\mathbf{D}_L^{(j)}\right\} + \left\{\mathbf{D}_T^{(j)}\right\} \quad (3.56)$$

where h_j denotes the thickness of the *j*th sublayer.

Considering the continuity of displacements and transverse stresses across the interfaces between these fictitious sublayers, we have

$$\left\{\mathbf{F}^{(j)}(h_j)\right\} = \left\{\mathbf{F}^{(j+1)}(0)\right\} \quad (3.57)$$

After establishing Equation (3.56) for all sublayers, the following equation can be obtained by using Equations (3.56) and (3.57) recursively:

$$\{\mathbf{F}(h_N)\} = [\mathbf{D}^{(N)}(h_N)]\{\mathbf{F}(h_{N-1})\} + \{\mathbf{D}_L^{(N)}\} + \{\mathbf{D}_T^{(N)}\}$$

$$= [\mathbf{D}^{(N)}(h_N)][[\mathbf{D}^{(N-1)}(h_{N-1})]\{\mathbf{F}(h_{N-2})\} + \{\mathbf{D}_L^{(N-1)}\} + \{\mathbf{D}_T^{(N-1)}\}] + \{\mathbf{D}_L^{(N)}\} + \{\mathbf{D}_T^{(N)}\}$$

$$= [\mathbf{D}^{(N)}(h_N)][\mathbf{D}^{(N-1)}(h_{N-1})]\{\mathbf{F}(h_{N-2})\}$$

$$\quad + [\mathbf{D}^{(N)}(h_N)]\{\{\mathbf{D}_L^{(N-1)}\} + \{\mathbf{D}_T^{(N-1)}\}\} + \{\mathbf{D}_L^{(N)}\} + \{\mathbf{D}_T^{(N)}\}$$

$$= \cdots\cdots$$

$$= [\mathbf{D}^{(N)}(h_N)][\mathbf{D}^{(N-1)}(h_{N-1})][\mathbf{D}^{(N-2)}(h_{N-2})]\cdots[\mathbf{D}^{(N-j)}(h_{N-j})]\{\mathbf{F}(h_{N-j-1})\}$$

$$\quad + [\mathbf{D}^{(N)}(h_N)][\mathbf{D}^{(N-1)}(h_{N-1})]\cdots[\mathbf{D}^{N-j+1}(h_{N-j+1})]\{\{\mathbf{D}_L^{(N-j)}\} + \{\mathbf{D}_T^{(N-j)}\}\}$$

$$\quad + [\mathbf{D}^{(N)}(h_N)][\mathbf{D}^{(N-1)}(h_{N-1})]\cdots[\mathbf{D}^{N-j+2}(h_{N-j+2})]\{\{\mathbf{D}_L^{(N-j+1)}\} + \{\mathbf{D}_T^{(N-j+1)}\}\} + \cdots$$

$$\quad + [\mathbf{D}^{(N)}(h_N)]\{\{\mathbf{D}_L^{(N-1)}\} + \{\mathbf{D}_T^{(N-1)}\}\} + \{\mathbf{D}_L^{(N)}\} + \{\mathbf{D}_T^{(N)}\}$$

$$= \cdots\cdots$$

$$= [\psi]\{\mathbf{F}(0)\} + \{\Omega\} \quad (3.58)$$

where

$$[\psi] = \prod_{j=N}^{1} [\mathbf{D}^{(j)}(h_j)]$$

$$[\Omega] = \sum_{i=2}^{N} \left(\prod_{j=N}^{i} [\mathbf{D}^{(j)}(h_j)] \right) \{\{\{\mathbf{D}_L^{(N-1)}\} + \{\mathbf{D}_T^{(N-1)}\}\} + \{\{\mathbf{D}_L^{(N)}\} + \{\mathbf{D}_T^{(N)}\}\} \quad (3.59)$$

It can be seen that Equation (3.58) has the same structure and dimension as in Equation (3.53). After introducing the boundary condition imposed on the two transverse surfaces and considering Equation (3.55), the surface displacements and/or stresses can be obtained. Introducing these solutions back into the equations at sublayer level, the displacements, stresses, and then strains within each sublayer can be further calculated.

3.5 Internal Surface Pressure Induced by a Medullar Pin

Prosthetic devices often employ metallic pins fitted into the medulla of a long bone as a means of attachment. These medullar pins cause the bone in the vicinity of the pin to change its internal structure and external shape. In this section we introduce the model presented in references 4, 9, and 14 for external changes in bone shape. The theory is applied here to the problem of determining the changes in external bone shape that result from a pin force-fitted into the medulla. The diaphysial region of a long bone is modeled here as a hollow circular cylinder, and external changes in shape are changes in the external and internal radii of the hollow circular cylinder.

The solution of this problem can be obtained by decomposing the problem into two separate subproblems: the problem of the remodeling of a hollow circular cylinder of adaptive bone material subjected to external loads, and the problem of an isotropic solid elastic cylinder subjected to an external pressure. These two problems are illustrated in Figure 3.2.

For an isotropic solid elastic cylinder subjected to an external pressure $p(t)$, the displacement in the radial direction is given by

$$u = \frac{-(2\mu + \lambda)p(t)r}{2\mu(3\lambda + 2\mu)} \quad (3.60)$$

where λ and μ are Lamé's constants for an isotropic solid elastic cylinder.

In this problem we calculate the pressure of interaction $p(t)$, which occurs when an isotropic solid cylinder of radius $a_0 + \delta/2$ is forced into a hollow adaptive bone cylinder of radius a_0.

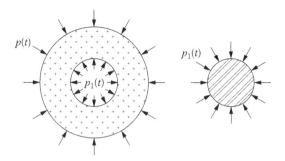

FIGURE 3.2
Decomposition of the medullar pin problem into two separate subproblems.

Let a and b denote the inner and outer radii, respectively, of the hollow bone cylinder at the instant after the solid isotropic cylinder has been forced into the hollow cylinder. Although the radii of the hollow cylinder will actually change during the adaptation process, the deviation of these quantities from a and b will be a small quantity negligible in small strain theory.

At an arbitrary instant in time after the two cylinders have been forced together, the pressure of the interaction is $p_1(t)$. The radial displacement of the solid cylinder at its surface is

$$u_1 = \frac{-(2\mu + \lambda)p_1(t)a}{2\mu(3\lambda + 2\mu)} \tag{3.61}$$

Using expression (3.29), the radial displacement of the bone at its inner surface is obtained as

$$u_2 = \frac{a}{F_3^*}\left(c_{33}\beta_1^*[\beta_2^* T_0 - p_1(t) + p(t)] + \varpi \frac{c_{33}c_{12}}{c_{11}} + \frac{F_2^* T_0 + P(t)}{\pi(b^2 - a^2)}c_{13} - F_1^* T_0 c_{13} \right)$$

$$+ \frac{a\beta_1^*[\beta_2^* T_0 - p_1(t) + p(t)]}{(c_{11} - c_{12})} - \frac{\varpi a}{c_{11}} \tag{3.62}$$

As it is assumed that the two surfaces have perfect contact, the two displacements have the relationship

$$a_0 + \frac{\delta}{2} + u_1 = a_0 + u_2 \tag{3.63}$$

Hence, we find

$$\delta = 2(u_2 - u_1) \tag{3.64}$$

Substituting Equations (3.61) and (3.62) into Equation (3.63) and then solving Equation (3.63) for $p_1(t)$, we obtain

$$p_1(t) = -\frac{1}{H}\left[\frac{\delta}{a} - \left(H_1\frac{b^2}{b^2-a^2} + H_2\frac{1}{b^2-a^2} + H_3\frac{1}{\ln b/a}\right)\right] \qquad (3.65)$$

where

$$H = \frac{2\mu+\lambda}{\mu(3\lambda+2\mu)} - 2\left(\frac{c_{33}}{F_3^*} + \frac{1}{c_{11}-c_{12}}\right)\frac{b^2}{b^2-a^2} \qquad (3.66)$$

$$H_1 = \frac{2}{F_3^*}\left[c_{13}\left(\frac{c_{13}}{c_{11}}\lambda_1 - \lambda_3\right)T_0\right] - 2\left(\frac{c_{33}}{F_3^*} + \frac{1}{c_{11}-c_{12}}\right)\left[\beta_2^*T_0 + p(t)\right] \qquad (3.67)$$

$$H_2 = \frac{2}{F_3^*}\frac{c_{13}P(t)}{\pi} \qquad (3.68)$$

$$H_3 = \frac{c_{33}c_{12}\lambda_1 T_0}{F_3^* c_{11}} - \frac{2}{F_3^*}\left(\frac{c_{13}}{c_{11}}\lambda_1 - \frac{\lambda_3}{2}\right)c_{13}T_0 - \frac{\lambda_1 T_0}{c_{11}} \qquad (3.69)$$

Equation (3.65) is the solution of the internal surface pressure induced by an inserting medullar pin.

3.6 Numerical Examples

As numerical illustration of the analytical and semianalytical solutions described before, a femur with $a = 25$ mm and $b = 35$ mm is considered. The material properties assumed for the bone are

$$c_{11} = 15(1+e)\text{GPa}, \quad c_{12} = c_{13} = 6.6(1+e)\text{GPa}, \quad c_{33} = 12(1+e)\text{GPa},$$

$$c_{44} = 4.4(1+e)\text{GPa}, \quad \lambda_1 = 0.621(1+e) \times 10^5\text{NK}^{-1}\text{m}^{-2},$$

$$\lambda_3 = 0.551(1+e) \times 10^5\text{NK}^{-1}\text{m}^{-2}, \quad \chi_3 = 0.0133(1+e)\text{CK}^{-1}\text{m}^{-2}$$

$$e_{31} = -0.435(1+e)\text{C/m}^2, \quad e_{33} = 1.75(1+e)\text{C/m}^2, \qquad (3.70)$$

$$e_{15} = 1.14(1+e)\text{C/m}^2, \quad \kappa_1 = 111.5(1+e)\kappa_0,$$

$$\kappa_3 = 126(1+e)\kappa_0, \quad \kappa_0 = 8.85 \times 10^{-12}\text{C}^2/\text{Nm}^2 = \text{permittivity of free space}$$

The remodeling rate coefficients are assumed to be

$$C_0 = 3.09 \times 10^{-9}\text{sec}^{-1}, \quad C_1 = 2 \times 10^{-7}\text{sec}^{-1}, \quad C_2 = 10^{-6}\text{sec}^{-1}$$

and

$$A_{rr}^{\varepsilon 0} = A_{rr}^{\varepsilon 1} = A_{zz}^{\varepsilon 0} = A_{zz}^{\varepsilon 1} = A_{rz}^{\varepsilon 0} = A_{rz}^{\varepsilon 1} = 10^{-5}\,\text{sec}^{-1},$$

$$A_{r}^{E0} = A_{r}^{E1} = 10^{-15}\,\text{V}^{-1}\text{m/sec} = 10^{-15}\,\text{N}^{-1}\text{C/sec}$$

The initial inner and outer radii are assumed to be

$$a_0 = 25\,\text{mm}, \quad b_0 = 35\,\text{mm}$$

and $e_0 = 0$ is assumed. In the calculation, $u_r(t) \ll a_0$ is assumed for the sake of simplicity; that is, $a(t)$ and $b(t)$ may be approximated by a_0 and b_0.

3.6.1 A Hollow, Homogeneous Circular Cylindrical Bone Subjected to Various External Loads

To analyze remodeling behavior affected by various loading cases we distinguish the following five loading cases:

- Case 1: $p(t) = n \times 2$ MPa ($n = 1, 2, 3,$ and 4), $P(t) = 1500$ N, with no other type of load applied
  - Table 3.1 lists the results at some typical instances of time, obtained by both the analytical and the semianalytical solutions. The semianalytical solution is obtained by dividing the bone into N (= 10, 20, 40) sublayers. It is evident from the table, and also from other extensive comparisons that are not shown here, that the solutions have excellent agreement on the change rate of porosity e. Hence, for the numerical results presented later, no references are given regarding which method is used to obtain the solution, unless otherwise stated. It is also evident from the table that the numerical results will gradually converge to the exact value as the layer number N increases.
  - The extended results for this loading case are shown in Figure 3.3, demonstrating the effect of external pressure on the bone remodeling process. It is evident that there is a critical value, p_{r0}, above

TABLE 3.1

Comparison of Porosity e Obtained by Analytical and Semianalytical Solutions ($P = 1500$ N, $p = 2$ Mpa)

Time (sec)		500,000	1,000,000	1,500,000	2,000,000
Semianalytical	$N = 10$	7.283×10^{-5}	1.533×10^{-4}	2.423×10^{-4}	3.406×10^{-4}
	$N = 20$	7.294×10^{-5}	1.536×10^{-4}	2.427×10^{-4}	3.412×10^{-4}
	$N = 40$	7.297×10^{-5}	1.536×10^{-4}	2.428×10^{-4}	3.413×10^{-4}
Analytical		7.298×10^{-5}	1.536×10^{-4}	2.428×10^{-4}	3.414×10^{-4}

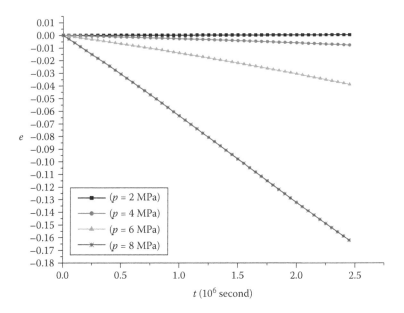

FIGURE 3.3
Variation of e with time t ($\varphi_b - \varphi_a = T_0 = 0$ and $P = 1500$ N).

which the porosity of the femur will be reduced. The critical value p_{r0} in this problem is approximately 2.95 MPa. It is also evident that the porosity of the femur increases along with the increase of external pressure p.

- Case 2: $P = 1500$ N and internal pressure is produced by inserting a rigid pin whose radius a^* is greater than a

  - The values of e as a function of t for $a^* - a = 0.01, 0.03,$ and 0.05 mm are shown in Figure 3.4. It is interesting to note that for the three cases, the bone structure at the pin–bone interface adapts itself initially to become less porous and then to a state with even less porosity. This is followed by a quick recovery of porosity, indicated by a sharply decreased value of e. As time approaches infinity, the bone structure stabilizes itself at a moderately reduced porosity. Although dramatic change of the remodeling constant is observed during the remodeling process, it is believed that the effect of the change on bone structures is limited by the fact that the duration of the change is very short compared to the entire remodeling process. This result coincides with Cowin and Van Buskirk's [4] theoretical observation, which showed that a bone structure might tend to a physiologically impossible bone structure in finite time. Both these conditions have been observed clinically and classified as osteoporosis (excess density with the

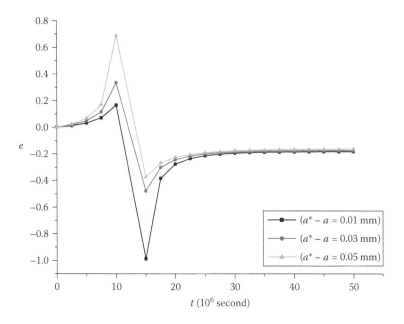

FIGURE 3.4
Variation of e with time induced by a solid pin.

maximum value of e) and osteopetrosis (excess porosity with the minimum value of e), respectively.

- Figure 3.4 also shows the variation of e against tightness of fit. It is evident that the tightness of fit has significant effects on the remodeling process, especially during the period when the abrupt change of porosity occurs. It must be mentioned here that the remodeling rate for this period can only serve as an indication of the modeling process, since Equation (3.4) is only valid for predicting a low remodeling rate. Thus, detailed analysis of the equation will not provide any further reliable information. More sophisticated and advanced remodeling models are evidently needed. Nevertheless, the prediction does suggest that the possibility exists of loss of grip on the pin or of high-level tensile stresses in the bone layer surrounding the pin, which may induce cracks.
- Case 3: $T_0(t) = 10°C, 20°C, 30°C, 40°C,$
  - Figure 3.5 shows the effects of temperature change on bone remodeling rate at $r = b_0$ when $\varphi_b - \varphi_a = p(t) = P(t) = 0$. In general, low temperature induces more porous bone structures, whereas a warmer environment may improve the remodeling process with a less porous bone structure. After considering all other factors, it is expected that there is a preferred temperature under which an ideal remodeling rate may be achieved.

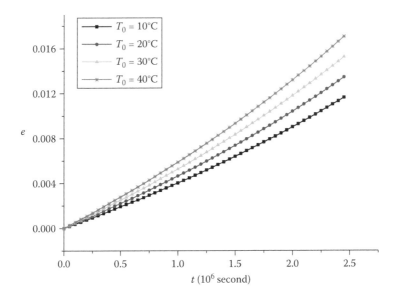

FIGURE 3.5
Variation of e with time t for several temperatures ($\varphi_b - \varphi_a = p = P = 0$).

- Case 4: $\varphi_b - \varphi_a = -60, -30, 30,$ and 60 V, $r = b_0$, and $T_0 = p = P = 0$
  - Figure 3.6 shows the variation of e with time t for various values of electric potential difference with $T_0 = p = P = 0$. It can be observed from Figure 3.6 that there are no significant differences between the remodeling rates when the external electric potential difference $\varphi_b - \varphi_a$ changes from -60 to 60 V, though it is evident that the remodeling rate increases as the electric potential difference decreases. However, the result does suggest that the remodeling process might be improved by exposing a bone to an electric field. Further theoretical and experimental studies are needed to investigate the implication of this in medical practice.
- Case 5: $\varphi_b - \varphi_a = -60, -30, 30,$ and 60 V, $p(t) = 2$ MPa, $P(t) = 1500$ N, and $T_0 = 0$
  - This loading case is considered in order to study the coupling effect of electrical and mechanical loads on bone remodeling rate. Figure 3.7 shows the numerical results of volume fraction change against different values of electric potential difference $\varphi_b - \varphi_a$ when $T_0 = 0$, $P(t) = 1500$ N, and $p(t) = 2$ MPa. As already observed in Figure 3.6, it can be seen from Figure 3.7 that the bone remodeling rate increases along with the decrease of the potential difference $\varphi_b - \varphi_a$. The combination of electrical and mechanical loads results in significantly different values of the remodeling rate when different electrical fields are applied.

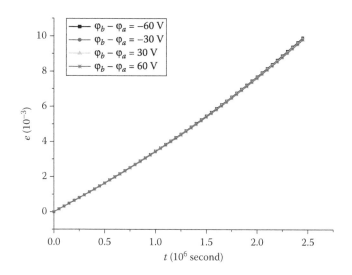

FIGURE 3.6
Variation of e with time t for several potential differences ($T_0 = p = P = 0$).

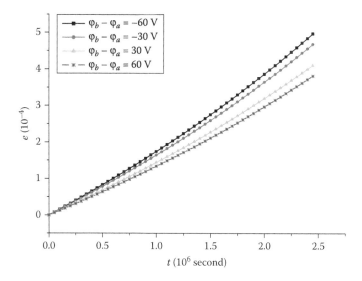

FIGURE 3.7
Variation of e with time t for coupling loads ($p = 2$ MPa, $P = 1500$ N, and $T_0 = 0$).

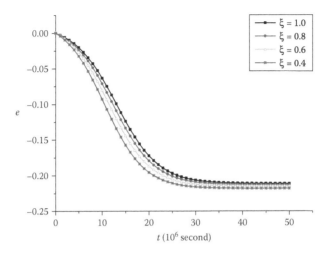

FIGURE 3.8
Variation of e with time t for an inhomogeneous bone subjected to coupling loads (p = 4 MPa, P = 1500 N, T_0 = 40, and $\varphi_b - \varphi_a$ = 30V).

3.6.2 A Hollow, Inhomogeneous Circular Cylindrical Bone Subjected to External Loads

The geometrical and material parameters of this problem are the same as those used in the preceding cases, except that all material constants in Equation (3.70) are now modified by a multiplier $[1 - (1 - \xi)(b - r)/(b - a)]$, where $0 \le \xi \le 1$ and represents the percentage reduction of stiffness at the inner surface of the bone. It is worth mentioning that by using the semianalytical approach, the form of stiffness variation in the radial direction can be arbitrary. Figure 3.8 shows the results of e at the outside surface of the bone for ξ = 1, 0.8, 0.6, and 0.4. The external loads are p = 4 MPa, P = 1500 N, T_0 = 40, and $\varphi_b - \varphi_a = 30$ V. In general, the remodeling rate declines as the initial stiffness of the inner bone surface decreases. When time approaches infinity, it is observed that the stiffness reduction in the radial direction has an insignificant effect on the remodeling rate of the outside bone surface. This observation suggests that ignoring stiffness reduction in the radial direction can yield satisfactory prediction of the remodeling process occurring at the outside layer of the bone.

3.7 Extension to Thermomagnetoelectroelastic Solid

In the case of a thermomagnetoelectroelastic solid, the constitutive relation (3.1) and the remodeling rate equation (3.4) must be augmented by some additional terms related to magnetic field as follows [11]:

$$\sigma_{rr} = c_{11}\varepsilon_{rr} + c_{12}\varepsilon_{\theta\theta} + c_{13}\varepsilon_{zz} - e_{31}E_z - \alpha_{31}H_z - \lambda_1 T,$$

$$\sigma_{\theta\theta} = c_{12}\varepsilon_{rr} + c_{11}\varepsilon_{\theta\theta} + c_{13}\varepsilon_{zz} - e_{31}E_z - \alpha_{31}H_z - \lambda_1 T,$$

$$\sigma_{zz} = c_{13}\varepsilon_{rr} + c_{13}\varepsilon_{\theta\theta} + c_{33}\varepsilon_{zz} - e_{33}E_z - \alpha_{33}H_z - \lambda_3 T,$$

$$\sigma_{zr} = c_{44}\varepsilon_{zr} - e_{15}E_r - \alpha_{15}H_r,$$

$$D_r = e_{15}\varepsilon_{zr} + \kappa_1 E_r + d_1 H_r,$$ (3.71)

$$D_z = e_{31}(\varepsilon_{rr} + \varepsilon_{\theta\theta}) + e_{33}\varepsilon_{zz} + \kappa_3 E_z + d_3 H_z - \chi_3 T,$$

$$B_r = \alpha_{15}\varepsilon_{zr} + d_1 E_r + \mu_1 H_r,$$

$$B_z = \alpha_{31}(\varepsilon_{rr} + \varepsilon_{\theta\theta}) + \alpha_{33}\varepsilon_{zz} + d_3 E_z + \mu_3 H_z - m_3 T,$$

$$h_r = k_r q_r, \qquad h_z = k_z q_z$$

$$\dot{e} = A(e) + A_r^E(e)E_r + A_z^E(e)E_z + A_r^M(e)H_r + A_z^M(e)H_z$$

$$+ A_{rr}^\varepsilon(e)(\varepsilon_{rr} + \varepsilon_{\theta\theta}) + A_{zz}^\varepsilon(e)\varepsilon_{zz} + A_{rz}^\varepsilon(e)\varepsilon_{rz}$$ (3.72)

where

B_i and H_i are components of magnetic induction and magnetic field, respectively

α_{ij} are piezomagnetic constants

d_i are magnetoelectric constants

μ_i are magnetic permeabilities

m_3 is the pyromagnetic constant

The associated magnetic field is related to magnetic potential ψ by

$$H_r = \psi_{,r} \quad H_z = \psi_{,z}$$ (3.73)

If we consider again a hollow circular cylinder of bone as shown in Figure 3.1, subjected to an external temperature change T_0, a quasistatic axial load P, an external pressure p, an electric potential load φ_a (or/and φ_b), and a magnetic potential load ψ_a (and/or ψ_b), the governing equations (3.8) and (3.9) and boundary condition (3.5) are, in this case, replaced by

$$\left(\frac{\partial^2}{\partial r^2} + \frac{1}{r}\frac{\partial}{\partial r} \right)(c_{44}u_z + e_{15}\varphi + \alpha_{15}\psi) = 0$$ (3.74)

$$\left(\frac{\partial^2}{\partial r^2} + \frac{1}{r}\frac{\partial}{\partial r} \right)(e_{15}u_z - \kappa_1\varphi - d_1\psi) = 0$$ (3.75)

$$\left(\frac{\partial^2}{\partial r^2} + \frac{1}{r}\frac{\partial}{\partial r} \right)(\alpha_{15}u_z - d_1\varphi - \mu_1\psi) = 0$$ (3.76)

$$T = 0, \quad \sigma_{rr} = \sigma_{r\theta} = \sigma_{rz} = 0, \quad \varphi = \varphi_a, \quad \psi = \psi_a \quad \text{at } r = a,$$

$$T = T_0, \quad \sigma_{rr} = -p, \quad \sigma_{r\theta} = \sigma_{rz} = 0, \quad \varphi = \varphi_b, \quad \psi = \psi_b \quad \text{at } r = b \tag{3.77}$$

The solutions to the governing equations (3.7), (3.74), and (3.75) satisfying boundary conditions (3.6) and (3.77) are given by

$$u_r = \frac{r}{F_3^*}\left(c_{33}\beta_1^*[\beta_2^* T_0 + p(t)] + \varpi \frac{c_{33}c_{12}}{c_{11}} + \frac{F_2^* T_0 + P(t)}{\pi(b^2 - a^2)} c_{13} - F_1^* T_0 c_{13} \right)$$
$$+ \frac{a^2 \beta_1^*[\beta_2^* T_0 + p(t)]}{r(c_{11} - c_{12})} + \frac{\varpi r[\ln(r/a) - 1]}{c_{11}} \tag{3.78}$$

$$u_z = \frac{z}{F_3^*}\left(\left[F_1^* T_0 - \frac{F_2^* T_0 + P(t)}{\pi(b^2 - a^2)} \right](c_{11} + c_{12}) - 2c_{13}\beta_1^*[\beta_2^* T_0 + p(t)] \right.$$
$$\left. - \frac{2c_{13}c_{12}\varpi}{c_{11}} \right) - \frac{e_{15}(\varphi_b - \varphi_a)\ln(r/a)}{c_{44}\ln(b/a)} - \frac{\alpha_{15}(\psi_b - \psi_a)\ln(r/a)}{c_{44}\ln(b/a)} \tag{3.79}$$

$$\varphi = \frac{\ln(r/a)}{\ln(b/a)}(\varphi_b - \varphi_a) + \varphi_a, \quad \psi = \frac{\ln(r/a)}{\ln(b/a)}(\psi_b - \psi_a) + \psi_a \tag{3.80}$$

$$T = \frac{\ln(r/a)}{\ln(b/a)} T_0 \tag{3.81}$$

The strains, electric, and magnetic field intensity can be found by introducing Equations (3.78)–(3.81) into Equation (3.71). They are, respectively,

$$\varepsilon_{rr} = \frac{1}{F_3^*}\left(c_{33}\beta_1^*[\beta_2^* T_0 + p(t)] + \varpi \frac{c_{33}c_{12}}{c_{11}} + \frac{F_2^* T_0 + P(t)}{\pi(b^2 - a^2)} c_{13} - F_1^* T_0 c_{13} \right)$$
$$- \frac{a^2 \beta_1^*[\beta_2^* T_0 + p(t)]}{r^2(c_{11} - c_{12})} + \frac{\varpi \ln(r/a)}{c_{11}} \tag{3.82}$$

$$\varepsilon_{\theta\theta} = \frac{1}{F_3^*}\left(c_{33}\beta_1^*[\beta_2^* T_0 + p(t)] + \varpi \frac{c_{33}c_{12}}{c_{11}} + \frac{F_2^* T_0 + P(t)}{\pi(b^2 - a^2)} c_{13} - F_1^* T_0 c_{13} \right)$$
$$+ \frac{a^2 \beta_1^*[\beta_2^* T_0 + p(t)]}{r^2(c_{11} - c_{12})} + \frac{\varpi[\ln(r/a) - 1]}{c_{11}} \tag{3.83}$$

$$\varepsilon_{zz} = \frac{1}{F_3^*}\left(\left[F_1^* T_0 - \frac{F_2^* T_0 + P(t)}{\pi(b^2 - a^2)} \right](c_{11} + c_{12}) - 2c_{13}\beta_1^*[\beta_2^* T_0 + p(t)] - \frac{2c_{13}c_{12}\varpi}{c_{11}} \right) \tag{3.84}$$

$$\varepsilon_{rz} = -\frac{e_{15}(\varphi_b - \varphi_a)}{rc_{44}\ln(b/a)} - \frac{\alpha_{15}(\psi_b - \psi_a)}{rc_{44}\ln(b/a)} \tag{3.85}$$

$$E_r = -\frac{(\varphi_b - \varphi_a)}{r\ln(b/a)}, \quad H_r = -\frac{(\psi_b - \psi_a)}{r\ln(b/a)} \tag{3.86}$$

Substituting Equation (3.82)–(3.86) yields

$$\dot{e} = A^*(e) + \frac{2A_{rr}^s}{F_3^*}\left(c_{33}\beta_1^*[\beta_2^*T_0 + p(t)] + \varpi\frac{c_{33}c_{12}}{c_{11}} + \frac{F_2^*T_0 + P(t)}{\pi(b^2 - a^2)}c_{13} - F_1^*T_0c_{13}\right)$$

$$+ \frac{A_{rr}^s\varpi[2\ln(r/a) - 1]}{c_{11}} + \frac{A_{zz}^s}{F_3^*}\left(\left[F_1^*T_0 - \frac{F_2^*T_0 + P(t)}{\pi(b^2 - a^2)}\right](c_{11} + c_{12})\right.$$

$$\left.- 2c_{13}\beta_1^*[\beta_2^*T_0 + p(t)] - \frac{2c_{13}c_{12}\varpi}{c_{11}}\right) - \frac{\phi_b - \phi_a}{r\ln(b/a)}\left(A_r^E + \frac{e_{15}}{c_{44}}A_{zr}^s\right) \tag{3.87}$$

$$\underline{- \frac{\psi_b - \psi_a}{r\ln(b/a)}\left(G_r^E + \frac{\alpha_{15}}{c_{44}}A_{zr}^s\right)}$$

In comparison with Equation (3.37), the only difference is due to the last term (underlined term).

As numerical illustration of the bone devolution process, the results presented in Qu and Qin [11] are summarized here. In these authors' work, a femur with $a = 25$ mm and $b = 35$ mm is considered. The material properties assumed for the bone are

$$c_{11} = 15(1+e)\text{GPa}, \quad c_{12} = c_{13} = 6.6(1+e)\text{GPa}, \quad c_{33} = 12(1=e)\text{GPa},$$

$$c_{44} = 4.4(1+e)\text{GPa}, \quad \lambda_1 = 0.621(1+e) \times 10^5\text{NK}^{-1}\text{m}^{-2},$$

$$\lambda_3 = 0.55(1+e) \times 10^5\text{NK}^{-1}\text{m}^{-2}, \quad \chi_3 = 0.133(1+e)\text{CK}^{-1}\text{m}^{-2} \tag{3.88}$$

$$e_{31} = -0.435(1+e)\text{C/m}^2, \quad e_{33} = 1.75(1+e)\text{C/m}^2,$$

$$e_{15} = 1.14(1+e)\text{C/m}^2, \quad \kappa_1 = 111.5(1+e)\kappa_0, \quad \alpha_{15} = 550(1+e)\text{N/Am}$$

$$\kappa_3 = 126(1+e)\kappa_0, \quad \kappa_0 = 8.85 \times 10^{-12}\text{C}^2/\text{Nm}^2 = \text{permittivity of free space}$$

The remodeling rate coefficients are assumed to be

$$C_0 = 3.09 \times 10^{-9}\text{sec}^{-1}, \quad C_1 = 2 \times 10^{-7}\text{sec}^{-1}, \quad C_2 = 10^{-6}\text{sec}^{-1}$$

and

$$A_{rr}^{\varepsilon 0} = A_{rr}^{\varepsilon 1} = A_{zz}^{\varepsilon 0} = A_{zz}^{\varepsilon 1} = A_{rz}^{\varepsilon 0} = A_{rz}^{\varepsilon 1} = 10^{-5} \text{ sec}^{-1},$$

$$A_r^{E0} = A_r^{E1} = 10^{-15} \text{V}^{-1}\text{m}/\text{sec} = 10^{-15} \text{N}^{-1}\text{C}/\text{sec}$$

$$G_r^{E0} = G_r^{E1} = 1.5 \times 10^{-8} \text{m}/(\text{Ampere}\cdot\text{sec})$$

The initial inner and outer radii are assumed to be

$$a_0 = 25\,\text{mm}, \quad b_0 = 35\,\text{mm}$$

and $e_0 = 0$ is assumed. In the calculation, $u_r(t) \ll a_0$ has again been assumed for the sake of simplicity; that is, $a(t)$ and $b(t)$ may be approximated by a_0 and b_0.

To illustrate the devolution process, we investigate the change of the volume fraction of bone matrix material from its reference value, which is denoted by e, in the transverse direction at several specific times. We also distinguish three loading cases to investigate the influence of electric, magnetic, and thermal loads on the bone structure. Finally, the effect on the bone of coupling electric and mechanical loads is studied.

- Case 1: $p(t) = 0$, $P = 1500$ N, $T_0(t) = 0°$C, $\varphi_b - \varphi_a = 30$V, $\psi_b - \psi_a = 0$.
  - Figure 3.9 shows the variation of e with time t in the transverse direction of bone for the electric loading described previously. It can be seen from the figure that as time passes, the initially

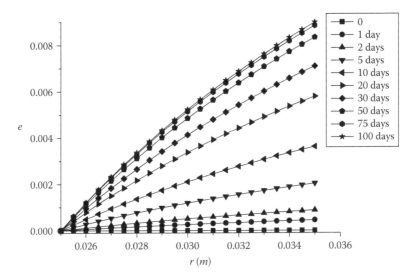

FIGURE 3.9
Variation of e with time t along the radii for electric load.

homogeneous bone structure gradually becomes inhomogeneous. The change in the volume fraction of bone matrix material on its inner surface is less than that on its outer surface. This means that the bone tissue near the outer surface is less porous and thus denser than that near the inner surface, which means that it is stronger. This can be illustrated by the theory of adaptive elasticity. After the transverse electric field is loaded, an inhomogeneous stress field is generated. Then the stress of the inner surface is smaller than that of the outer one. As the bone remodeling process is ongoing, the strain field becomes more homogeneous. To achieve this, the bone tissue must change to a state with more porous endosteum and less porous periosteum, which results in an inhomogeneous bone structure. Although the value of e is very small, transverse electric loads can indeed change the bone structure. If real remodeling rate coefficients are obtained by way of experiment, then we can evaluate the effect of electric field on the bone structure. It is also found that as time approaches infinity, the value of e gradually decreases. This indicates that the bone structure stabilizes itself at a relatively steady state, which can be accepted as the end of the remodeling process.

- Case 2: $p(t) = 0$, $P = 1500$ N, $T_0(t) = 0°C$, $\varphi_b - \varphi_a = 0$, $\psi_b - \psi_a = 1A$.

  - Figure 3.10 shows the corresponding variation of e with time t along the radii of bone for the magnetic loading case. It can be

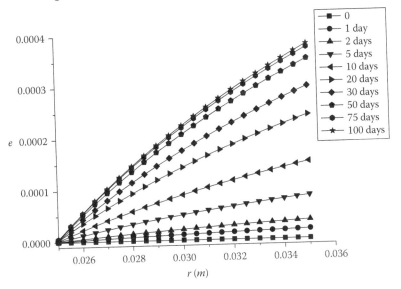

FIGURE 3.10
Variation of e with time t along the radii for magnetic load.

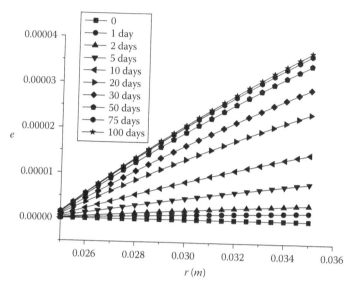

FIGURE 3.11
Variation of *e* with time *t* along the radii for thermal load.

seen from the figure that a magnetic load has a similar influence on bone structure to an electric load. A magnetic load can also inhomogenize an initially homogeneous bone structure through the bone remodeling process. Further experimental and theoretical investigations need to be developed to obtain the exact remodeling rate coefficients and to discover the importance of the role played by magnetic stimuli.

- Case 3: $p(t) = 0$, $P = 1500$ N, $T_0(t) = 0.1°C$, $\varphi_b - \varphi_a = 0$, $\psi_b - \psi_a = 0$.
  - Figure 3.11 shows the variation of *e* with time *t* along the radii of bone when the thermal loading case is applied. A similar phenomenon to that of Figure 3.10 is evident in Figure 3.11, indicating that a warmer environment may improve the remodeling process with a less porous bone structure, and change of temperature can also result in an inhomogeneous bone structure. As mentioned in Qin and Ye [19], the process by which temperature change can affect bone remodeling is still an open question. An initial purpose of the study presented in Qu and Qin [11] was to show how a bone may respond to thermal, magnetic, and electric loads and to provide information for the possible use of imposed external temperature and/or magnetic-electrical fields in medical treatment and in controlling the healing process of injured bones. Further investigations are undoubtedly needed.

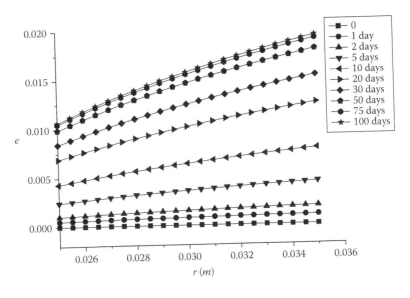

FIGURE 3.12
Variation of e with time t along the radii for coupled loads.

- Case 4: $p(t) = 1$ MPa, $P = 1500$ N, $T_0(t) = 0.1°C$, $\varphi_b - \varphi_a = 30V$, $\psi_b - \psi_a = 1A$.
  - Figure 3.12 shows the variation of e with time t in the transverse direction when subjected to coupled loads. The preceding loading case is considered to study the coupling effect of electric-magnetic and mechanical loads on bone structure. It can be seen from Figure 3.12 that the function of coupled loads is the superposition of the constituent single loads. They are, however, not simply linearly superposed. The combination of magnetic, electric, thermal, and mechanical loads results in significant changes in bone structure and properties of bone tissues. This indicates that loading coupled fields is more effective in modifying bone structure than loading only one kind of field.

References

1. Frost H. M. Dynamics of bone remodeling. In *Bone biodynamics*, ed. Frost H. M., p. 316. Boston: Little and Brown (1964).
2. Charnay A., Tschantz J. Mechanical influences in bone remodeling, experimental research on Wolff's law. *Journal of Biomechanics* 5: 173–180 (1972).
3. Cowin S. C., Hegedus D. H. Bone remodeling.1. Theory of adaptive elasticity. *Journal of Elasticity* 6 (3): 313–326 (1976).

4. Cowin S. C., Van Buskirk W. C. Internal bone remodeling induced by a medullary pin. *Journal of Biomechanics* 11 (5): 269–275 (1978).

5. Cowin S. C., Van Buskirk W. C. Surface bone remodeling induced by a medullary pin. *Journal of Biomechanics* 12 (4): 269–276 (1979).

6. Cowin S. C., Firoozbakhsh K. Bone remodeling of diaphyseal surfaces under constant load—Theoretical predictions. *Journal of Biomechanics* 14 (7): 471–484 (1981).

7. Hart R. T., Davy D. T., Heiple K. G. A computational method for stress analysis of adaptive elastic materials with a view toward applications in strain-induced bone remodeling. *Journal of Biomechanical Engineering—Transactions of the ASME* 106 (4): 342–350 (1984).

8. Tsili M. C. Theoretical solutions for internal bone remodeling of diaphyseal shafts using adaptive elasticity theory. *Journal of Biomechanics* 33 (2): 235–239 (2000).

9. Qin Q. H., Qu C. Y., Ye J. Q. Thermo electroelastic solutions for surface bone remodeling under axial and transverse loads. *Biomaterials* 26 (33): 6798–6810 (2005).

10. Qu C. Y., Qin Q. H., Kang Y. L. A hypothetical mechanism of bone remodeling and modeling under electromagnetic loads. *Biomaterials* 27 (21): 4050–4057 (2006).

11. Qu C. Y., Qin Q. H. Evolution of bone structure under axial and transverse loads. *Structural Engineering and Mechanics* 24 (1): 19–29 (2006).

12. He X. Q., Qu C., Qin Q. H. A theoretical model for surface bone remodeling under electromagnetic loads. *Archive of Applied Mechanics* 78 (3): 163–175 (2008).

13. Qu C. Y., Qin Q. H. Bone remodeling under multifield coupled loading. *CAAI Transactions on Intelligent Systems* 2 (3): 52–58 (2007).

14. Qu C. Y., Qin Q. H., Kang Y. L. Thermomagnetoelectroelastic prediction of the bone surface remodeling under axial and transverse loads. In *9th International Conference on Inspection, Appraisal, Repairs & Maintenance of Structures*, ed. Ren W. X., Ong K. C. G., Tan J. S. Y., Fuzhou, China, October 20–21 2005, pp. 373–380. CI-Premier PTE LTD.

15. Fukada E., Yasuda I. On the piezoelectric effect of bone. *Journal of the Physical Society of Japan* 12 (10): 1158–1162 (1957).

16. Fukada E., Yasuda I. Piezoelectric effects in collagen. *Japanese Journal of Applied Physics* 3 (2): 117–121 (1964).

17. Kryszewski M. Fifty years of study of the piezoelectric properties of macromolecular structured biological materials. *Acta Physica Polonica A* 105 (4): 389–408 (2004).

18. Robiony M., Polini F., Vercellotti T., Politi M. Piezoelectric bone cutting in multipiece maxillary osteotomies. *Journal of Oral and Maxillofacial Surgery* 62 (6): 759–761 (2004).

19. Qin Q. H., Ye J. Q. Thermoelectroelastic solutions for internal bone remodeling under axial and transverse loads. *International Journal of Solids and Structures* 41 (9–10): 2447–2460 (2004).

20. Gjelsvik A. Bone remodeling and piezoelectricity 1. *Journal of Biomechanics* 6 (1): 69–77 (1973).

21. Williams W. S., Breger L. Analysis of stress distribution and piezoelectric response in cantilever bending of bone and tendon. *Annals of the New York Academy of Sciences* 238 (OCT11): 121–130 (1974).

22. Guzelsu N. Piezoelectric model for dry bone tissue. *Journal of Biomechanics* 11 (5): 257–267 (1978).
23. Johnson M. W., Williams W. S., Gross D. Ceramic models for piezoelectricity in dry bone. *Journal of Biomechanics* 13 (7): 565–573 (1980).
24. Demiray H. Electro-mechanical remodeling of bones. *International Journal of Engineering Science* 21 (9): 1117–1126 (1983).
25. Aschero G., Gizdulich P., Mango F., Romano S. M. Converse piezoelectric effect detected in fresh cow femur bone. *Journal of Biomechanics* 29 (9): 1169–1174 (1996).
26. Aschero G., Gizdulich P., Mango F. Statistical characterization of piezoelectric coefficient d(23) in cow bone. *Journal of Biomechanics* 32 (6): 573–577 (1999).
27. Fotiadis D. I., Foutsitzi G., Massalas C. V. Wave propagation in human long bones of arbitrary cross section. *International Journal of Engineering Science* 38 (14): 1553–1591 (2000).
28. El-Naggar A. M., Abd-Alla A. M., Mahmoud S. R. Analytical solution of electro-mechanical wave propagation in long bones. *Applied Mathematics and Computation* 119 (1): 77–98 (2001).
29. Ahmed S. M., Abd-Alla A. M. Electromechanical wave propagation in a cylindrical poroelastic bone with cavity. *Applied Mathematics and Computation* 133 (2–3): 257–286 (2002).
30. Silva C. C., Thomazini D., Pinheiro A. G., Aranha N., Figueiro S. D., Goes J. C., Sombra A. S. B. Collagen-hydroxyapatite films: Piezoelectric properties. *Materials Science and Engineering B—Solid State Materials for Advanced Technology* 86 (3): 210–218 (2001).
31. Qin Q. H. Multi-field bone remodeling under axial and transverse loads. In *New research on biomaterials,* ed. Boomington D. R., pp. 49–91. New York: Nova Science Publishers (2007).
32. Qu C. Y. Simulation of bone remodeling under multi-field loadings. PhD thesis, Tianjin University (2007).
33. Qin Q. H. *Fracture mechanics of piezoelectric materials.* Southampton, England: WIT Press (2001).
34. Hegedus D. H., Cowin S. C. Bone remodeling—II. Small strain adaptive elasticity. *Journal of Elasticity* 6 (4): 337–352 (1976).
35. Ye J. Q. *Laminated composite plates and shells: 3D modeling.* London: Springer-Verlag (2002).

4

Multifield Surface Bone Remodeling

4.1 Introduction

In Chapter 3, the theoretical and numerical results of internal bone remodeling were presented. Extension to multifield surface bone remodeling is discussed in this chapter. Theoretical predictions of surface bone remodeling in the diaphysis of the long bone under various external loads are made within the framework of adaptive elastic theory. These loads include external lateral pressure, and electric and thermal loads. Two solutions are presented for analyzing thermoelectroelastic problems of surface bone remodeling. The analytical solution that gives explicit formulation is capable of modeling homogeneous bone materials, and the semianalytical solution is suitable for analyzing inhomogeneous cases. Numerical results are presented to verify the proposed formulation and to show the effects of mechanical, thermal, and electric loads on the surface bone remodeling process. In this chapter, the developments in references 1–3 are briefly summarized.

4.2 Solution of Surface Modeling for a Homogeneous Hollow Circular Cylindrical Bone

In this section, analytical solutions for thermoelectroelastic problems of surface bone remodeling, based on the theory of adaptive elasticity [4], are presented to study the effects of mechanical, thermal, and electric loads on the surface bone remodeling process. The solution is used to investigate the surface bone remodeling process on the basis of assuming a homogeneous bone material [5].

4.2.1 Rate Equation for Surface Bone Remodeling

The equations for the theory of adaptive elasticity (2.1) can be extended to include piezoelectric effects by adding some new terms as follow [1,3]:

$$U = C_{ij}(n,Q)\left[\varepsilon_{ij}(Q,t) - \varepsilon_{ij}^0(Q,t)\right] + C_i(n,Q)\left[E_i(Q,t) - E_i^0(Q,t)\right]$$

$$= C_{rr}\varepsilon_{rr} + C_{\theta\theta}\varepsilon_{\theta\theta} + C_{zz}\varepsilon_{zz} + C_{rz}\varepsilon_{rz} + C_r E_r + C_z E_z - C_0 \tag{4.1}$$

where $C_0 = C_{rr}\varepsilon_{rr}^0 + C_{zz}\varepsilon_{zz}^0 + C_{\theta\theta}\varepsilon_{\theta\theta}^0 + C_{rz}\varepsilon_{rz}^0 + C_r E_r^0 + C_z E_z^0$, C_i are surface remodeling coefficients.

4.2.2 Differential Field Equation for Surface Remodeling

We now consider again the hollow circular cylinder of bone shown in Figure 3.1. The hollow circular cylinder is subjected to an external temperature change T_0, a quasistatic axial load P, an external pressure p, and an electric potential load φ_a(or/and φ_b), as shown in Figure 3.1. The boundary conditions are defined by Equations (3.5) and (3.6). The solution of displacements u_r, u_z and electric potential φ to the preceding problem were given by Equations (3.29)–(3.36) in Chapter 3. Substituting Equations (3.32)–(3.36) into Equation (4.1) yields

$$U_e = N_1^e \frac{b^2}{b^2 - a^2} + N_2^e \frac{1}{\ln\left(\dfrac{b}{a}\right)} + N_3^e \frac{1}{b^2 - a^2} + N_4^e \frac{1}{a\ln\left(\dfrac{b}{a}\right)} - C_0^e \tag{4.2}$$

$$U_p = N_1^p \frac{b^2}{b^2 - a^2} + N_1^{p'} \frac{a^2}{b^2 - a^2} + N_2^p \frac{1}{\ln\left(\dfrac{b}{a}\right)} + N_3^p \frac{1}{b^2 - a^2}$$

$$+ N_4^p \frac{1}{b\ln\left(\dfrac{b}{a}\right)} + N_3' - C_0^p \tag{4.3}$$

where

$$N_1^e = \frac{1}{F_3^*}\left\{c_{13}\left(\frac{c_{13}}{c_{11}}\lambda_1 - \lambda_3\right)T_0 - c_{33}\left[\frac{\lambda_1}{2}\left(\frac{c_{12}}{c_{11}} - 1\right)T_0 + p(t)\right]\right\}(C_{rr}^e + C_{\theta\theta}^e)$$

$$+ \frac{1}{F_3^*}\left\{2c_{13}C_{zz}^e\left[\frac{\lambda_1}{2}(\frac{c_{12}}{c_{11}} - 1)T_0 + p(t)\right] - (c_{11} + c_{12})C_{zz}^e\left(\frac{c_{13}}{c_{11}}\lambda_1 - \lambda_3\right)T_0\right\} \tag{4.4}$$

$$+ \frac{(C_{rr}^e - C_{\theta\theta}^e)\left[\dfrac{\lambda_1}{2}\left(\dfrac{c_{12}}{c_{11}} - 1\right)T_0 + p(t)\right]}{c_{11} - c_{12}}$$

$$N_2^e = \frac{1}{F_3^*}\left[\frac{c_{12}c_{33}}{2c_{11}}\lambda_1 T_0 - \left(\frac{c_{13}}{c_{11}}\lambda_1 - \frac{\lambda_3}{2}\right)c_{13}T_0\right]\left(C_{rr}^e + C_{\theta\theta}^e\right)$$

$$+ \frac{C_{zz}^e}{F_3^*}\left[(c_{11}+c_{12})\left(\frac{c_{13}}{c_{11}}\lambda_1 - \frac{\lambda_3}{2}\right)T_0 - \frac{c_{12}c_{13}}{c_{11}}\lambda_1 T_0\right] - \frac{C_{\theta\theta}^e \lambda_1 T_0}{2c_{11}}$$

(4.5)

$$N_3^e = \frac{1}{F_3^*}\left[c_{13}\left(C_{rr}^e + C_{\theta\theta}^e\right) - (c_{11}+c_{12})C_{zz}^e\right]\frac{P(t)}{\pi} \tag{4.6}$$

$$N_4^e = -\left(\frac{e_{15}}{c_{44}}C_{zr}^e + C_r\right)\left(\varphi_b - \varphi_a\right) \tag{4.7}$$

$$N_1^p = \frac{1}{F_3^*}\left\{c_{13}\left(\frac{c_{13}}{c_{11}}\lambda_1 - \lambda_3\right)T_0 - c_{33}\left[\frac{\lambda_1}{2}\left(\frac{c_{12}}{c_{11}}-1\right)T_0 + p(t)\right]\right\}\left(C_{rr}^p + C_{\theta\theta}^p\right)$$

$$+ \frac{1}{F_3^*}\left\{2c_{13}C_{zz}^p\left[\frac{\lambda_1}{2}\left(\frac{c_{12}}{c_{11}}-1\right)T_0 + p(t)\right] - (c_{11}+c_{12})C_{zz}^p\left(\frac{c_{13}}{c_{11}}\lambda_1 - \lambda_3\right)T_0\right\}$$

(4.8)

$$N_2^p = \frac{1}{F_3^*}\left[\frac{c_{12}c_{33}}{2c_{11}}\lambda_1 T_0 - \left(\frac{c_{13}}{c_{11}}\lambda_1 - \frac{\lambda_3}{2}\right)c_{13}T_0\right]\left(C_{rr}^p + C_{\theta\theta}^p\right)$$

$$+ \frac{C_{zz}^p}{F_3^*}\left[(c_{11}+c_{12})\left(\frac{c_{13}}{c_{11}}\lambda_1 - \frac{\lambda_3}{2}\right)T_0 - \frac{c_{12}c_{13}}{c_{11}}\lambda_1 T_0\right] - \frac{C_{\theta\theta}^p \lambda_1 T_0}{2c_{11}}$$

(4.9)

$$N_3^p = \frac{1}{F_3^*}\left[c_{13}\left(C_{rr}^p + C_{\theta\theta}^p\right) - (c_{11}+c_{12})C_{zz}^p\right]\frac{P(t)}{\pi} \tag{4.10}$$

$$N_4^p = -\left(\frac{e_{15}}{c_{44}}C_{zr}^p + C_r\right)\left(\varphi_b - \varphi_a\right) \tag{4.11}$$

$$N_1^{p'} = -\frac{(C_{rr}^p - C_{\theta\theta}^p)\left[\frac{\lambda_1}{2}\left(\frac{c_{12}}{c_{11}}-1\right)T_0 + p(t)\right]}{c_{11}-c_{12}} \tag{4.12}$$

$$N_3' = \frac{\left(C_{\theta\theta}^p + C_{rr}^p\right)\lambda_1 T_0}{2c_{11}} \tag{4.13}$$

and the subscripts (or superscripts) p and e refer to periosteal and endosteal, respectively. For the reason stated in Section 2.3.2, Equation (2.21) can be used here—that is,

$$U_e = -\frac{da}{dt}, \quad U_p = \frac{db}{dt} \tag{4.14}$$

Thus, Equations (4.2) and (4.3) can be written as

$$-\frac{da}{dt} = N_1^e \frac{b^2}{b^2 - a^2} + N_2^e \frac{1}{\ln\left(\dfrac{b}{a}\right)} + N_3^e \frac{1}{b^2 - a^2} + N_4^e \frac{1}{a\ln\left(\dfrac{b}{a}\right)} - C_0^e \quad (4.15)$$

$$\frac{db}{dt} = N_1^p \frac{b^2}{b^2 - a^2} + N_1^{p'} \frac{a^2}{b^2 - a^2} + N_2^p \frac{1}{\ln\left(\dfrac{b}{a}\right)} + N_3^p \frac{1}{b^2 - a^2} + N_4^p \frac{1}{b\ln\left(\dfrac{b}{a}\right)} - C_0^{p'} \quad (4.16)$$

where
$$C_0^{p'} = C_0^p - N_3'.$$

4.2.3 Approximation for Small Changes in Radii

It is apparent that Equations (4.15) and (4.16) are nonlinear and cannot, in general, be solved analytically. However, the equations can be approximately linearized when they are applied to solve problems with small changes in radii. In the bone surface remodeling process, Qin, Qu, and Ye [1] assumed that the radii of the inner and outer surface of the bone change very little compared to their original values, as was done in Section 2.3.2. This means that the changes in $a(t)$ and $b(t)$ are small. This is a reasonable assumption from the viewpoint of the physics of the problem. To introduce the approximation, two nondimensional parameters [3,4],

$$\varepsilon = \frac{a}{a_0} - 1, \quad \eta = \frac{b}{b_0} - 1, \quad (4.17)$$

are adopted in the following calculations. As a result, $a(t)$ and $b(t)$ can be written as

$$a(t) = (1 + \varepsilon(t))a_0, \ b(t) = (1 + \eta(t))b_0, \ (\varepsilon, \eta \ll 1) \quad (4.18)$$

Since both ε and η are far smaller than one, their squares can be ignored from the equations. Consequently, we can have the following approximations:

$$\frac{b^2}{b^2 - a^2} \cong L_0 + 2L_0^2 \frac{a_0^2}{b_0^2}(\varepsilon - \eta) \quad (4.19)$$

$$\frac{a^2}{b^2 - a^2} \cong L_0' + 2L_0^{2'} \frac{b_0^2}{a_0^2}(\varepsilon - \eta) \quad (4.20)$$

$$\frac{1}{b^2 - a^2} \cong L_2 + 2L_2^2(a_0^2\varepsilon - b_0^2\eta) \quad (4.21)$$

$$\frac{1}{\ln\left(\dfrac{b}{a}\right)} \cong L_1 + L_1^2(\varepsilon - \eta) \tag{4.22}$$

$$\frac{1}{a\ln\left(\dfrac{b}{a}\right)} \cong \frac{1}{a_0}L_1(1-\varepsilon) + \frac{1}{a_0}L_1^2(\varepsilon - \eta) \tag{4.23}$$

$$\frac{1}{b\ln\left(\dfrac{b}{a}\right)} \cong \frac{1}{b_0}L_1(1-\eta) + \frac{1}{b_0}L_1^2(\varepsilon - \eta) \tag{4.24}$$

where

$$L_0 = \frac{b_0^2}{b_0^2 - a_0^2}, \quad L_0' = \frac{a_0^2}{b_0^2 - a_0^2}, \quad L_1 = \frac{1}{\ln\left(\dfrac{b_0}{a_0}\right)}, \quad L_2 = \frac{1}{b_0^2 - a_0^2} \tag{4.25}$$

Thus, Equations (4.15) and (4.16) can be approximately represented in terms of ε and η as follows:

$$\frac{d\varepsilon}{dt} = B_1\varepsilon + B_2\eta + B_3, \qquad \frac{d\eta}{dt} = B_1'\varepsilon + B_2'\eta + B_3' \tag{4.26}$$

where

$$B_1 = -\frac{1}{a_0}\left(2L_0^2\frac{a_0^2}{b_0^2}N_1^e + L_1^2N_2^e - \frac{L_1N_4^e}{a_0} + \frac{L_1^2N_4^e}{a_0} + 2L_2^2a_0^2N_3^e\right) \tag{4.27}$$

$$B_2 = \frac{1}{a_0}\left(2L_0^2\frac{a_0^2}{b_0^2}N_1^e + L_1^2N_2^e + \frac{L_1^2N_4^e}{a_0} + 2L_2^2b_0^2N_3^e\right) \tag{4.28}$$

$$B_3 = -\frac{1}{a_0}\left(L_0N_1^e + L_1N_2^e + L_2N_3^e + \frac{L_1N_4^e}{a_0} - C_0^e\right) \tag{4.29}$$

$$B_1' = \frac{1}{b_0}\left(2L_0^2\frac{a_0^2}{b_0^2}N_1^p + L_1^2N_2^p + \frac{L_1^2N_4^p}{b_0} + 2L_2^2a_0^2N_3^p + 2L_0^2\frac{b_0^2}{a_0^2}N_1^{p'}\right) \tag{4.30}$$

$$B_2' = -\frac{1}{b_0}\left(2L_0^2\frac{a_0^2}{b_0^2}N_1^p + L_1^2N_2^p - \frac{L_1N_4^p}{b_0} + \frac{L_1^2N_4^p}{b_0} + 2L_2^2b_0^2N_3^p + 2L_0^2\frac{b_0^2}{a_0^2}N_1^{p'}\right) \tag{4.31}$$

$$B_3' = \frac{1}{b_0}\left(L_0N_1^p + L_1N_2^p + L_2N_3^p + \frac{L_1N_4^p}{b_0} + L_0'N_1^{p'} - C_0^{p'}\right) \tag{4.32}$$

4.2.4 Analytical Solution of Surface Remodeling

An analytical solution of Equation (4.26) can be obtained if a homogeneous property is assumed for bone materials. In that case, the inhomogeneous linear differential equations system (4.26) can be converted into the following homogeneous one:

$$\frac{d\varepsilon'}{dt} = B_1\varepsilon' + B_2\eta', \quad \frac{d\eta'}{dt} = B_1'\varepsilon' + B_2'\eta' \tag{4.33}$$

by introducing two new variables such that

$$\varepsilon' = \varepsilon - \varepsilon_\infty, \quad \eta' = \eta - \eta_\infty \tag{4.34}$$

$$\varepsilon_\infty = \frac{1}{\det\mathbf{M}}(B_3'B_2 - B_3B_2'), \quad \eta_\infty = \frac{1}{\det\mathbf{M}}(B_3B_1' - B_3'B_1) \tag{4.35}$$

$$\mathbf{M} = \begin{pmatrix} B_1 & B_2 \\ B_1' & B_2' \end{pmatrix} \tag{4.36}$$

$$\det\mathbf{M} = B_1B_2' - B_1'B_2 \tag{4.37}$$

The solution of Equation (4.33), subject to the initial conditions that $\varepsilon(0) = 0$ and $\eta(0) = 0$, can be expressed in four possible forms that fulfill the physics of the problem (i.e., when $t \to \infty$, ε and η must be limited quantities, $a < b$, and the solution must be stable). The form of the solution depends on the roots of the following quadratic equation:

$$s^2 - tr\mathbf{M}s + \det\mathbf{M} = 0 \tag{4.38}$$

where

$$tr\mathbf{M} = B_1 + B_2' = s_1 + s_2 \tag{4.39}$$

All the theoretically possible solutions are shown as follows:

Case A: When $(B_1 - B_2')^2 + 4B_2B_1' > 0$, $B_1 + B_2' < 0$, and $B_1B_2' - B_2B_1' > 0$, Equation (4.38) has two different roots, s_1 and s_2, both of which are real and distinct. Then the solutions of the equations are

$$\varepsilon' = \frac{1}{s_1 - s_2}\left[(s_2\varepsilon_\infty - B_3)e^{-s_1t} + (B_3 - s_1\varepsilon_\infty)e^{-s_2t}\right],$$

$$\eta' = \frac{1}{s_1 - s_2}\left[(s_2\eta_\infty - B_3')e^{-s_1t} + (B_3' - s_1\eta_\infty)e^{-s_2t}\right] \tag{4.40}$$

This can also be written as

$$\varepsilon(t) = \varepsilon_\infty + \frac{1}{s_1 - s_2}\left[\left(s_2\varepsilon_\infty - B_3\right)e^{-s_1 t} + \left(B_3 - s_1\varepsilon_\infty\right)e^{-s_2 t}\right],$$

$$\eta(t) = \eta_\infty + \frac{1}{s_1 - s_2}\left[\left(s_2\eta_\infty - B_3'\right)e^{-s_1 t} + \left(B_3' - s_1\eta_\infty\right)e^{-s_2 t}\right] \tag{4.41}$$

The formulae for the variation of the radii—that is, $a(t)$ and $b(t)$—with time t can be obtained by substituting Equation (4.41) into Equation (4.17). Thus,

$$a(t) = a_0 + a_0\varepsilon_\infty + \frac{a_0}{s_1 - s_2}\left[\left(s_2\varepsilon_\infty - B_3\right)e^{-s_1 t} + \left(B_3 - s_1\varepsilon_\infty\right)e^{-s_2 t}\right],$$

$$b(t) = b_0 + b_0\eta_\infty + \frac{b_0}{s_1 - s_2}\left[\left(s_2\eta_\infty - B_3'\right)e^{-s_1 t} + \left(B_3' - s_1\eta_\infty\right)e^{-s_2 t}\right] \tag{4.42}$$

The final radii of the cylinder are then

$$a_\infty = \lim_{t\to\infty} a(t) = a_0(1 + \varepsilon_\infty), \quad b_\infty = \lim_{t\to\infty} b(t) = b_0(1 + \eta_\infty) \tag{4.43}$$

Case B: When $(B_1 - B_2')^2 + 4B_2 B_1' = 0,\quad B_1 \neq B_2',$ and $B_1 + B_2' < 0,$ Equation (4.38) has two equal roots, $B_2' + B_1$. The solutions of the equations are

$$\varepsilon' = -\left\{\varepsilon_\infty + \left[\frac{B_1 - B_2'}{2}\varepsilon_\infty + B_2\eta_\infty\right]t\right\}e^{\frac{B_2' + B_1}{2}t}$$

$$\eta' = -\left\{\eta_\infty - \left[\frac{(B_2' - B_1)^2}{4B_2}\varepsilon_\infty - \frac{B_2' - B_1}{2}\eta_\infty\right]t\right\}e^{\frac{B_2' + B_1}{2}t} \tag{4.44}$$

This can also be written as

$$\varepsilon(t) = \varepsilon_\infty - \left\{\varepsilon_\infty + \left[\frac{B_1 - B_2'}{2}\varepsilon_\infty + B_2\eta_\infty\right]t\right\}e^{\frac{B_2' + B_1}{2}t},$$

$$\eta(t) = \eta_\infty - \left\{\eta_\infty - \left[\frac{(B_2' - B_1)^2}{4B_2}\varepsilon_\infty - \frac{B_2' - B_1}{2}\eta_\infty\right]t\right\}e^{\frac{B_2' + B_1}{2}t} \tag{4.45}$$

The formulae for the variation of $a(t)$ and $b(t)$ with time t can be obtained by substituting Equation (4.45) into Equation (4.17) as

$$a(t) = a_0 + a_0\varepsilon_\infty - a_0 \left\{ \varepsilon_\infty + \left[\frac{B_1 - B_2'}{2} \varepsilon_\infty + B_2\eta_\infty \right] t \right\} e^{\frac{B_2' + B_1}{2} t},$$

$$b(t) = b_0 + b_0\eta_\infty - b_0 \left\{ \eta_\infty - \left[\frac{(B_2' - B_1)^2}{4B_2} \varepsilon_\infty - \frac{B_2' - B_1}{2} \eta_\infty \right] t \right\} e^{\frac{B_2' + B_1}{2} t}$$

$$(4.46)$$

The final radii of the cylinder are then

$$a_\infty = \lim_{t \to \infty} a(t) = a_0(1 + \varepsilon_\infty), \quad b_\infty = \lim_{t \to \infty} b(t) = b_0(1 + \eta_\infty) \tag{4.47}$$

Case C: When $B_1 = B_2' < 0$ and $B_2 = 0$, the solutions of the equations are

$$\varepsilon' = -\varepsilon_\infty e^{B_1 t}, \quad \eta' = -\left(B_1' \varepsilon_\infty t + \eta_\infty \right) e^{B_1 t} \tag{4.48}$$

This can also be written as

$$\varepsilon(t) = \varepsilon_\infty - \varepsilon_\infty e^{B_1 t}, \quad \eta(t) = \eta_\infty - \left(B_1' \varepsilon_\infty t + \eta_\infty \right) e^{B_1 t} \tag{4.49}$$

The formulae for the variation of $a(t)$ and $b(t)$ with time t can be obtained by substituting Equation (4.49) into Equation (4.17), as follows:

$$a(t) = a_0 + a_0\varepsilon_\infty \left(1 - e^{B_1 t} \right), \quad b(t) = b_0 + b_0\eta_\infty - b_0 \left(B_1'\varepsilon_\infty t + \eta_\infty \right) e^{B_1 t} \tag{4.50}$$

The final radii of the cylinder are then

$$a_\infty = \lim_{t \to \infty} a(t) = a_0(1 + \varepsilon_\infty), \quad b_\infty = \lim_{t \to \infty} b(t) = b_0(1 + \eta_\infty) \tag{4.51}$$

Case D: When $B_1 = B_2' < 0$ and $B_1' = 0$, the solutions of the equations are

$$\varepsilon' = -\left(B_2\eta_\infty t + \varepsilon_\infty \right) e^{B_1 t}, \quad \eta' = -\eta_\infty e^{B_1 t} \tag{4.52}$$

This can also be written as

$$\varepsilon(t) = \varepsilon_\infty - \left(B_2\eta_\infty t + \varepsilon_\infty \right) e^{B_1 t}, \quad \eta(t) = \eta_\infty - \eta_\infty e^{B_1 t} \tag{4.53}$$

The formulae for the variation of the radii with time t can be obtained by substituting Equation (4.53) into Equation (4.17). Thus,

$$a(t) = a_0 + a_0\varepsilon_\infty - a_0\left(B_2\eta_\infty t + \varepsilon_\infty\right)e^{B_1 t}$$

$$b(t) = b_0 + b_0\eta_\infty\left(1 - e^{B_1 t}\right) \tag{4.54}$$

The final radii of the cylinder are then

$$a_\infty = \lim_{t\to\infty} a(t) = a_0(1+\varepsilon_\infty), \quad b_\infty = \lim_{t\to\infty} b(t) = b_0(1+\eta_\infty) \tag{4.55}$$

All of these solutions are theoretically valid. However, the first is the most likely solution to the problem, as it is physically possible when $t\to\infty$ [1]. Therefore, it can be used to calculate the bone surface remodeling.

4.3 Application of Semianalytical Solution to Surface Remodeling of Inhomogeneous Bone

The semianalytical solution presented in Section 3.4 can be used to calculate strains and stresses at any point on the bone surface. These results form the basis for surface bone remodeling analysis. This section presents applications of solution (3.58) to the analysis of surface remodeling behavior in inhomogeneous bone.

It is noted that surface bone remodeling is a time-dependent process. The change in the radii (ε or η) can therefore be calculated by using the rectangular algorithm of integral (see Figure 4.1). The procedure is described here.

First, let T_0 be the starting time and T be the length of time to be considered and divide the time domain T into m equal intervals $\Delta T = T/m$. At the time t, calculate the strain and electric field using Equations (3.32)–(3.36). The results are then substituted into Equation (4.1) to determine the normal rate of the surface bone remodeling. Assuming that ΔT is sufficiently small, we can replace U with its mean value $\bar{U}$ at each time interval $[t, t+\Delta T]$. The change in the radii (ε or η) at time t can thus be determined using the results of surface velocity. Accordingly, the strain and electric field are updated by considering the change in the radii. The updated strain and electric field are in turn used to calculate the normal surface velocity at the next time interval. This process is repeated up to the last time interval $[T_0 + (m-1)\Delta T, T_0 + T]$. Figure 4.1 shows the rectangular algorithm of integral when we replace U with its initial value U_t (rather than its mean value $\bar{U}$) at the time interval $[t, t+\Delta T]$.

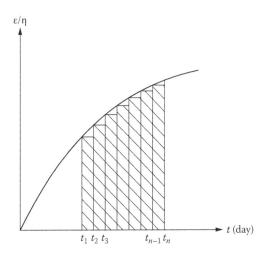

FIGURE 4.1
Illustration of the rectangular algorithm.

4.4 Surface Remodeling Equation Modified by an Inserting Medullar Pin

Substituting Equation (3.65) into Equations (4.15) and (4.16) yields

$$-\frac{da}{dt} = N_1^e \frac{b^2}{b^2 - a^2} + N_2^e \frac{1}{\ln\left(\frac{b}{a}\right)} + N_3^e \frac{1}{b^2 - a^2} + N_4^e \frac{1}{a\ln\left(\frac{b}{a}\right)}$$

$$- M_1 p_1(t) \frac{b^2}{a^2 - b^2} - C_0^e$$

(4.56)

$$\frac{db}{dt} = N_1^p \frac{b^2}{b^2 - a^2} + N_1^{p'} \frac{a^2}{b^2 - a^2} + N_2^p \frac{1}{\ln\left(\frac{b}{a}\right)} + N_3^p \frac{1}{b^2 - a^2}$$

$$- \left(M_2 \frac{b^2}{a^2 - b^2} + M_3 \frac{a^2}{a^2 - b^2} \right) p_1(t) + N_4^p \frac{1}{b\ln\left(\frac{b}{a}\right)} - C_0^{p'}$$

(4.57)

where

$$M_1 = \frac{(C_{rr}^e + C_{\theta\theta}^e)c_{33} - 2c_{13}C_{zz}^e}{F_3^*} - \frac{C_{rr}^e - C_{\theta\theta}^e}{c_{11} - c_{12}}$$

(4.58)

$$M_2 = \frac{(C_{rr}^p + C_{\theta\theta}^p)c_{33} - 2c_{13}C_{zz}^p}{F_3^*}$$

(4.59)

$$M_3 = -\frac{C_{rr}^p - C_{\theta\theta}^p}{c_{11} - c_{12}} \tag{4.60}$$

It can be seen that Equations (4.56) and (4.57) are similar to Equations (4.15) and (4.16). Considering Equation (4.17), Equations (4.56) and (4.57) can be simplified as

$$\frac{d\varepsilon}{dt} = Y_1 \varepsilon + Y_2 \eta + Y_3, \qquad \frac{d\eta}{dt} = Y_1' \varepsilon + Y_2' \eta + Y_3' \tag{4.61}$$

where

$$Y_1 = B_1 - M_1 \left[H_5 \Psi_1 + H_4 \left(2L_0^2 H_1 \frac{a_0^2}{b_0^2} + 2L_2^2 H_2 a_0^2 + H_3 L_1^2 + \frac{\delta}{a_0} \right) \right] \tag{4.62}$$

$$Y_2 = B_2 + M_1 \left[H_5 \Psi_1 + H_4 \left(2L_0^2 H_1 \frac{a_0^2}{b_0^2} + 2L_2^2 H_2 b_0^2 + H_3 L_1^2 \right) \right] \tag{4.63}$$

$$Y_3 = B_3 + M_1 H_4 \psi_1 \tag{4.64}$$

$$
\begin{aligned}
Y_1' = B_1' - M_2 &\left[H_5 \Psi_1 + H_4 \left(2L_0^2 H_1 \frac{a_0^2}{b_0^2} + 2L_2^2 H_2 a_0^2 + H_3 L_1^2 + \frac{\delta}{a_0} \right) \right] \\
&+ M_3 \left[H_7 \Psi_1 - H_6 \left(2L_0^2 H_1 \frac{a_0^2}{b_0^2} + 2L_2^2 H_2 a_0^2 + H_3 L_1^2 + \frac{\delta}{a_0} \right) \right]
\end{aligned} \tag{4.65}
$$

$$
\begin{aligned}
Y_2' = B_2' + M_2 &\left[H_5 \Psi_1 + H_4 \left(2L_0^2 H_1 \frac{a_0^2}{b_0^2} + 2L_2^2 H_2 b_0^2 + H_3 L_1^2 \right) \right] \\
&- M_3 \left[H_7 \Psi_1 - H_6 \left(2L_0^2 H_1 \frac{a_0^2}{b_0^2} + 2L_2^2 H_2 b_0^2 + H_3 L_1^2 \right) \right]
\end{aligned} \tag{4.66}
$$

$$Y_3' = B_3' + (M_2 H_4 + M_3 H_6) \Psi_1 \tag{4.67}$$

with

$$\Psi_1 = \frac{\delta}{a_0} - H_1 L_0 - H_2 L_2 - H_3 L_1 \tag{4.68}$$

$$H_4 = \frac{1}{2 \left(\dfrac{c_{33}}{F_3^*} + \dfrac{1}{c_{11} - c_{12}} \right) - \dfrac{2\mu + \lambda}{\mu(3\lambda + 2\mu)} \left(1 - \dfrac{a_0^2}{b_0^2} \right)} \tag{4.69}$$

$$H_5 = \frac{\dfrac{(4\mu + 2\lambda)a_0^2}{\mu(3\lambda + 2\mu)b_0^2}}{\left[2\left(\dfrac{c_{33}}{F_3^*} + \dfrac{1}{c_{11} - c_{12}}\right) - \dfrac{2\mu + \lambda}{\mu(3\lambda + 2\mu)}\left(1 - \dfrac{a_0^2}{b_0^2}\right)\right]^2} \tag{4.70}$$

$$H_6 = \frac{1}{\dfrac{2\mu + \lambda}{\mu(3\lambda + 2\mu)} + \left[2\left(\dfrac{c_{33}}{F_3^*} + \dfrac{1}{c_{11} - c_{12}}\right) - \dfrac{2\mu + \lambda}{\mu(3\lambda + 2\mu)}\dfrac{b_0^2}{a_0^2}\right]} \tag{4.71}$$

$$H_7 = \frac{2\left[2\left(\dfrac{c_{33}}{F_3^*} + \dfrac{1}{c_{11} - c_{12}}\right) - \dfrac{2\mu + \lambda}{\mu(3\lambda + 2\mu)}\dfrac{b_0^2}{a_0^2}\right]}{\left\{\dfrac{2\mu + \lambda}{\mu(3\lambda + 2\mu)} + \left[2\left(\dfrac{c_{33}}{F_3^*} + \dfrac{1}{c_{11} - c_{12}}\right) - \dfrac{2\mu + \lambda}{\mu(3\lambda + 2\mu)}\dfrac{b_0^2}{a_0^2}\right]\right\}^2} \tag{4.72}$$

Equation (4.61) is similar to Equation (4.33) and can thus be solved by following the solution procedure described in Section 4.2.4.

4.5 Numerical Examples for Thermopiezoelectric Bones

Consider again the femur used in Section 3.7. The geometrical and material coefficients of the femur are the same as those used in Section 3.6, except that the volume fraction change e is now taken to be zero here. In addition, the surface remodeling rate coefficients are assumed to be

$$C_{rr}^e = -9.6 \text{ m / sec}, \quad C_{\theta\theta}^e = -7.2 \text{ m / sec}, \quad C_{zz}^e = -5.4 \text{ m / sec},$$

$$C_{zr}^e = -8.4 \text{ m / sec}, \quad C_{rr}^p = -12.6 \text{ m / sec}, \quad C_{\theta\theta}^p = -10.8 \text{ m / sec},$$

$$C_{zz}^p = -9.6 \text{ m / sec}, \quad C_{zr}^p = -12 \text{ m / sec}, \quad C_r = 10^{-9} \text{ V}^{-1}\text{m}^2 \text{ / sec}$$

and

$$C_0^e = 0.0008373 \text{ m/sec}, \quad C_0^p = 0.00015843 \text{ m/sec}$$

and $\varepsilon_0 = 0$, $\eta_0 = 0$ are assumed.

In the following, numerical results are provided to show the effect of temperature and external electric load on the surface bone remodeling

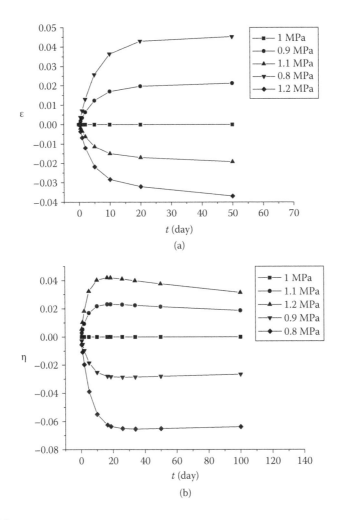

FIGURE 4.2
Variation of ε and η with time t ($P = 1500$ N; others $= 0$). (a) ε versus t; (b) η versus t.

process. The results for the effects of mechanical loading, inserted pin, and material inhomogeneity on the surface remodeling behavior are omitted here; they can be found in Qin et al. [1]. We distinguish five loading cases:

- Case 1: $p(t) = n \times 2$ MPa ($n = 0.8, 0.9, 1, 1.1$, and 1.2), $P(t) = 1500$ N, and no other type of load is applied
    - The extended results for this loading case are shown in Figure 4.2 to study the effect of external pressure on the bone remodeling

process. This figure illustrates that the transverse and axial loads have the same effect on the bone. The inner radius of the bone decreases while the outer radius increases as the external pressure increases. This results in an increase in the bone's cross-sectional area and, consequently, a thicker and stronger bone. When the external pressure decreases, the inner radius increases and the outer radius decreases; this means that the bone becomes thinner and weaker. On the other hand, greater pressure can increase the rate of bone surface remodeling, which can accelerate the recovery of injured bone.

- Figure 4.2 presents an interesting change of radius against time. The outer radius of the bone increases at first as the transverse pressure increases. It begins to decrease after a few days and finally converges to a stable value that is greater than its initial value. A similar result was presented by Cowin and Van Buskirk [4]. The earlier analysis was based on a model of surface remodeling and was attributed to the surrounding bone material becoming more or less stiff rather than being due to surface movement.

- Case 2: $T_0(t) = 29.5°C$, 29.8°C, 30°C, 30.2°C, 30.5°C, $\varphi_b - \varphi_a = 30V$, $p(t) = 1$ MPa, $P(t) = 1500$ N

  - Figure 4.3 shows the effects of temperature change on bone surface remodeling. In general, the radii of the bone decrease when the temperature increases and they increase when the temperature decreases. It can also be seen from the figure that ε and η are almost the same. Since $a_0 < b_0$, the change of the outer surface radius is normally greater than that of the inner one. The area of the bone cross section decreases as the temperature increases. This also suggests that a lower temperature is likely to induce thicker bone structures, whereas a warmer environment may improve the remodeling process with a less thick bone structure.

  - This result seems to coincide with actual fact. Thicker and stronger bones perhaps make a person living in Russia look stronger than one living in Vietnam. It should be mentioned here that how this change may affect the bone remodeling process is still an open question. As an initial investigation, the purpose of this study is to show how a bone may respond to thermal loads and to provide information for the possible use of imposed external temperature fields in medical treatment and in controlling the healing process of injured bones.

- Case 3: $\varphi_b - \varphi_a = -60$, −30, 30, and 60 V, $p(t) = 1$ MPa, $P(t) = 1500$ N, and $T_0 = 0$

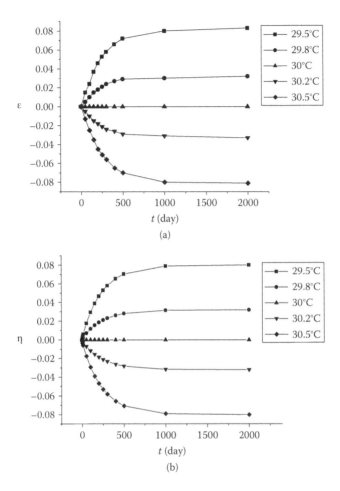

FIGURE 4.3
Variation of ε and η with time *t* for several temperature changes. (a) ε versus *t*; (b) η versus *t*.

- Figure 4.4 shows the variation of ε and η with time *t* for various values of electric potential difference. It can be seen that the effect of the electric potential is exactly opposite to that of temperature. A decrease in the intensity of the electric field results in a decrease of the inner and outer surface radii of the bone by almost the same magnitude. Theoretically, the results suggest that the remodeling process might be improved by exposing a bone to an electric field. Clearly, further theoretical and experimental studies are needed to investigate the implications of this exposure for medical practice.
- Case 4: $P = 1500$ N and internal pressure is produced by inserting a rigid pin whose radius a^* is greater than a

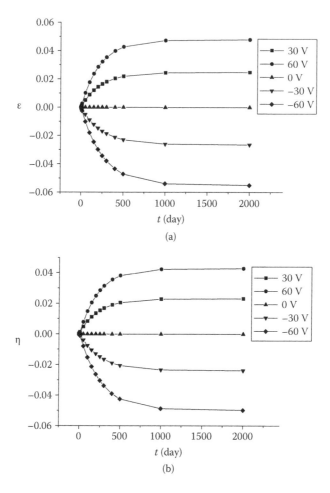

FIGURE 4.4
Variation of ε and η with time *t* for several potential differences. (a) ε versus *t*; (b) ε versus *t*.

- The values of ε and η against *t* for δ = $a^* - a$ = 0.001, 0.003, and 0.005 mm are shown in Figure 4.5. It can be seen that both the inner and the outer surface radii of the bone are reduced after the pin has been inserted, which will increase the tightness of fit. A tightly fitted pin can also increase the rate of bone surface remodeling, which can accelerate the recovery of the injured bone. On the other hand, as the radius of the pin increases, the outer surface radius of the bone decreases more significantly than does the inner surface radius. This results in a decrease of the bone's cross-sectional area and, hence, in a thinner and weaker bone structure. Furthermore, if the fit is too tight, the

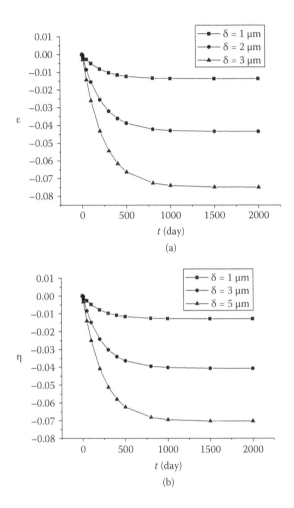

FIGURE 4.5
Variation of ε and η with time t induced by a solid pin. (a) ε versus t; (b) η versus t.

pressure on the interface will cause damage to the bone structure. Thus, the radius of the pin should be kept within an appropriate range.

- Case 5: A hollow, inhomogeneous circular cylindrical bone subjected to external loads
  - The external loads are $p = 1$ MPa, $P = 1500$ N, $T = 30°C$, and
  $\varphi_b - \varphi_a = 30$ V. The geometrical and material parameters of this
  problem are the same as those used in the previous cases, except
  that all material constants at the beginning of this section are now
  modified by a multiplier $[1 - (1 - \xi)(b - r)/(b - a)]$, where $0 \leq \xi \leq 1$

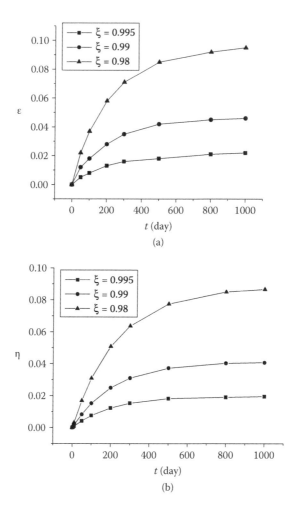

FIGURE 4.6

Variation of ε and η with time t for inhomogeneous bone subjected to coupling loads. (a) ε versus t; (b) η versus t.

and represents a percentage reduction of stiffness at the inner surface of the bone. It is worth mentioning that by using the semianalytical approach, the form of stiffness variation in the radial direction can be arbitrary. Figure 4.6 shows the results of ε and η at the outside surface of the bone for ξ = 0.995, 0.99, and 0.98. In general, the remodeling rate declines as the initial stiffness of the inner bone surface decreases. The inhomogeneity of the bone has a significant effect on the bone surface remodeling. A higher surface remodeling rate is always related to a higher level of inhomogeneity.

4.6 Extension to Thermomagnetoelectroelastic Solid

In the case of thermomagnetoelectroelastic solids, the constitutive equation is defined by Equation (3.71) and the remodeling rate equation (4.1) must be augmented by some additional terms related to magnetic field as follows:

$$U = C_{ij}(\mathbf{n}, Q)\left[\varepsilon_{ij}(Q) - \varepsilon_{ij}^0(Q)\right] + C_i\left[E_i(Q) - E_i^0(Q)\right] + G_i\left[H_i(Q) - H_i^0(Q)\right]$$

$$= C_{rr}\varepsilon_{rr} + C_{\theta\theta}\varepsilon_{\theta\theta} + C_{zz}\varepsilon_{zz} + C_{rz}\varepsilon_{rz} + C_r E_r + C_z E_z + G_r H_r + G_z H_z - C_0 \tag{4.73}$$

where G_i ($i = r,z$) are surface remodeling coefficients, and

$$C_0 = C_{rr}\varepsilon_{rr}^0 + C_{zz}\varepsilon_{zz}^0 + C_{\theta\theta}\varepsilon_{\theta\theta}^0 + C_{rz}\varepsilon_{rz}^0 + C_r E_r^0 + C_z E_z^0 + G_r H_r^0 + G_z H_z^0 \tag{4.74}$$

We now consider again a hollow circular cylinder of bone subjected to an external temperature change T_0, a quasistatic axial load P, an external pressure p, an electric potential load φ_a(or/and φ_b), and a magnetic potential load ψ_a(and/or ψ_b). The boundary conditions are defined by Equations (3.6) and (3.77). The governing equations are defined by Equations (3.7) and (3.74)–(3.76). Then, the solutions of U_e and U_p to the preceding problem have the same form as those of Equations (4.2) and (4.3), except that the coefficients N_4^e and N_4^p are replaced by

$$N_4^e = -\left(\frac{e_{15}}{c_{44}}C_{zr}^e + C_r\right)(\varphi_b - \varphi_a) - \left(\frac{\alpha_{15}}{c_{44}}C_{zr}^e + G_r\right)(\psi_b - \psi_a) \tag{4.75}$$

$$N_4^p = -\left(\frac{e_{15}}{c_{44}}C_{zr}^p + C_r\right)(\varphi_b - \varphi_a) - \left(\frac{\alpha_{15}}{c_{44}}C_{zr}^p + G_r\right)(\psi_b - \psi_a) \tag{4.76}$$

Therefore, the solutions to a and b have the same form as those described in Section 4.2.4.

As numerical illustration of the analytical solutions described in this section, we consider again a femur as discussed in Section 3.7, except that the volume fraction change e is now taken to be zero. In addition, the surface remodeling rate coefficients given in Section 4.5 are used. The surface remodeling rate coefficient for magnetic field is

$$G_r = 10^{-10}\text{m}^2/\text{Ampere}\cdot\text{sec}$$

We distinguish three loading cases:

- Case 1: $T_0(t) = -0.5°\text{C}, -0.2°\text{C}, 0°\text{C}, 0.2°\text{C}, 0.5°\text{C}, \varphi_b - \varphi_a = 30\text{V}, \psi_b - \psi_a = 1\text{A},$
  $p(t) = 1\text{ MPa}, P(t) = 1500\text{ N}$

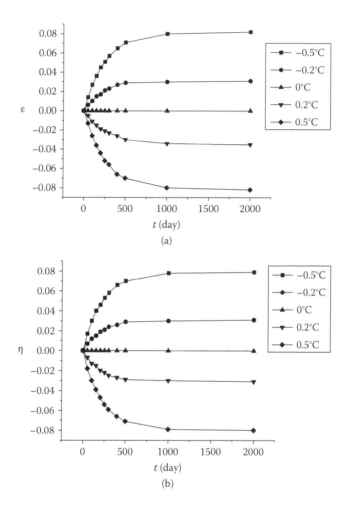

FIGURE 4.7
Variation of ε and η with time t for several temperature changes. (a) ε versus t; (b) η versus t.

- Similarly to that shown in Figure 4.3, Figure 4.7 shows the effects of temperature change on bone surface remodeling under the combination loading of mechanical, electric, and magnetic fields. The results obtained are similar to those of Figure 4.3 and the analysis for Figure 4.3 applies here.
- Case 2: $\varphi_b - \varphi_a = -60, -30, 30$, and 60 V, $p(t) = 1$ MPa, $P(t) = 1500$ N, $\psi_b - \psi_a = 1$A and $T_0 = 0$
  - Figure 4.8 shows the variation of ε and η with time t for various values of magnetic-field difference. Similarly to that shown in Figure 4.4, it can be also seen that the effect of the magnetic field

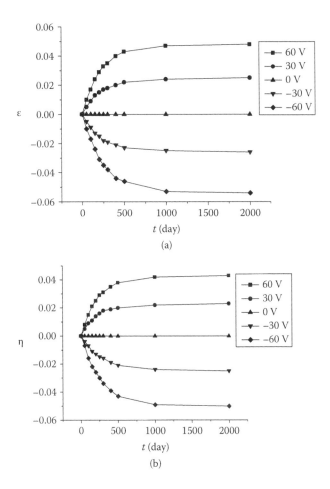

FIGURE 4.8
Variation of ε and η with time t for several electric potential differences. (a) ε versus t; (b) η versus t.

is exactly opposite to that of temperature. A decrease in the intensity of the magnetic field results in a decrease of the inner and outer surface radii of the bone by almost the same magnitude.

- Case 3: $\psi_b - \psi_a = -2, -1, 1$, and 2 A, $p(t) = 1$ MPa, $P(t) = 1500$ N, $\varphi_b - \varphi_a = 30$V and $T_0 = 0$
  - Figure 4.9 shows the variation of ε and η with time t for various values of magnetic potential difference. The changes of the outer and inner surfaces of the bone due to the magnetic influence are similar to those in Qin et al. [1].

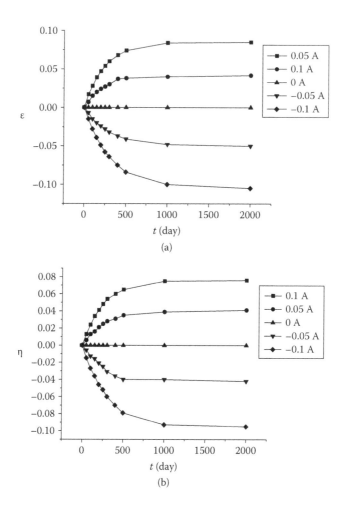

FIGURE 4.9
Variation of ε and η with time *t* for several magnetic potential differences. (a) ε versus *t*; (b) η versus *t*.

References

1. Qin Q. H., Qu C. Y., Ye J. Q. Thermo electroelastic solutions for surface bone remodeling under axial and transverse loads. *Biomaterials* 26 (33): 6798–6810 (2005).
2. Qu C. Y., Qin Q. H., Kang Y. L. Theoretical prediction of surface bone remodeling under axial and transverse loads. In *Proceedings of 9th International Conference on Inspection, Appraisal, Repairs & Maintenance of Structures,* ed. Ren W. X., Ong K. C. G., and Tan J. S. Y., pp. 373–380. CI-Premier PTY LTD (2005).

3. Qin Q. H. Multi-field bone remodeling under axial and transverse loads. In *New research on biomaterials,* ed. Boomington D. R., pp. 49–91. New York: Nova Science Publishers (2007).

4. Cowin S. C., Van Buskirk W. C. Surface bone remodeling induced by a medullary pin. *Journal of Biomechanics* 12 (4): 269–276 (1979).

5. Qin Q. H., Ye J. Q. Thermoelectroelastic solutions for internal bone remodeling under axial and transverse loads. *International Journal of Solids and Structures* 41 (9–10): 2447–2460 (2004).

5

Theoretical Models of Bone Modeling and Remodeling

5.1 Introduction

In the previous two chapters, internal and surface bone remodeling processes subjected to multifield loading were discussed. No effects of growth factors on bone remodeling were included in the models presented in those two chapters. In this chapter, a hypothetical regulation mechanism for bone modeling and remodeling under multifield loading presented in references 1–3 is discussed.

In this hypothesis, the bone modeling and remodeling mechanism is described as follows: The circular loads that we bear during ordinary daily activities generate microdamage in cortical bones and these microcracks are removed by osteoclasts. Then, growth factors, which are in a latent form in osteocytes, are activated by osteoclasts and released into bone fluids. These growth factors stimulate osteoblasts to refill the cavities. In particular, the related constitutive model presented in Hazelwood et al. [1] includes a number of relevant mechanical and biological processes and could be used to address differences in remodeling behavior as a volume element of bone is in a condition of disuse or overload. The model is then extended for analyzing surface bone remodeling under electromagnetic loading [3].

Functional adaptation of a living bone refers to the ability of the tissue to respond to changes in its environment. For cortical bone tissues, one potential response is remodeling, which involves a two-stage process carried out by teams of BMUs. Another response of cortical bone tissue is modeling, which refers to biological processes that produce functionally purposeful sizes and shapes of skeletal organs. For the most part, in a bone the processes involve modeling drifts of independent resorption and formation. Resorption of a packet of bone by osteoclasts is followed by refilling of the resorption cavity by osteoblasts. This sequence typically requires 3 to 4 months to complete at each locus, and the resorption and refilling cavities, while individually small, may collectively add substantial temporary porosity or "remodeling space" to the bone. The chief purpose of these processes seems to be to fit organs to their mechanical usage so that the usage does not break them or make them painful for a lifetime [4].

Before 1964, the literature did not distinguish between modeling and remodeling—lumping them together as remodeling. Since bones remodel themselves without the control of the nervous system, the most interesting feature of this process is that bone tissue seems to be capable of sensing the surrounding environment and controlling bone formation and resorption. Since the remodeling phenomenon was discovered by Wolff [5] in 1892, it has attracted widespread attention from biological scientists and mechanical engineering. Many hypotheses for this mechanism have been proposed, among which the theory of adaptive elasticity [6], the electricity theory [7,8], and the fatigue damage theory [9–11] are the most popular and widely used. Making use of these models, various computational models have been developed for testing the preceding hypotheses for the constitutive laws governing the mechanical adaptation of bones. Simulations have usually assumed that the functional stimulus for adaptation is stress, strain energy, electric field, magnetic field, or a related factor such as damage [1,3,10–17].

It should also be mentioned that in recent years much work has been done on the response of a bone tissue to an extremely low-frequency electromagnetic field [18–20]. Over the past two decades, increasing numbers of researchers have accepted the notion that there is a potential for interaction between electromagnetic fields and skeletal biological systems. Electromagnetic fields have been widely applied in the treatment of skeletal diseases such as osteoporosis, tendonitis, osteonecrosis, fracture, and nonunion. For example, electromagnetic stimulation of bone has been shown to promote osteogenesis in experiments using cultured cells [21,22] and in experiments in vivo using animals [23,24]. Moreover, extensive clinical experiments with electromagnetic stimulation for cases of nonunion or other osseous disease have revealed that electromagnetic stimulation can accelerate the recovery of those broken bone tissues [25,26]. All these results have shown that a pulsed electromagnetic field with extremely low frequency can trigger bone tissue to remodel itself. It is important, therefore, to investigate the mechanism of how an electromagnetic field acts on the functional adaptation of bone. With this background, Qu, Qin, and Kang [2] and He, Qu, and Qin [3] developed a theoretical model for analysis of bone modeling and remodeling under electromagnetic loading. In this chapter we focus on the developments in references 1–3.

5.2 Hypothetical Mechanism of Bone Remodeling

A hypothetical mechanism of internal and surface bone remodeling presented in references 2 and 3 is briefly reviewed in this section to provide reference for the remaining part of this chapter.

5.2.1 Bone Growth Factors

It has been reported that growth factors such as PDGF, IGF, bone morphogenetic protein (BMP), and TGF-β play an important role in bone formation and remodeling [27,28]. These growth factors are found in considerable quantities in bone matrix. Normally, they are retained in osteocytes. Once the osteocytes are resorbed, the growth factors can be released into bone fluid and can stimulate osteoblasts to refill resorption cavities. Unconsumed growth factors remaining in the bone fluid can be transported to the bone surface to deposit new bone material on it if the strain-induced fluid flow is strong enough to overcome the resistance. Experiments have shown that a pulsed, extremely low-frequency electromagnetic field can stimulate the multiplication of growth factors [29–31], thus indirectly accelerating the remodeling process via growth factors.

5.2.2 Electrical Signals in Bone Remodeling

Since 1957, when some bone tissues were found to have a piezoelectric effect [32], the electric properties of bone material have been widely investigated. It is believed that electrical signals in bone tissue play an important role in the bone modeling and remodeling processes [2,7,8,12,33]. The electrical signals that allow bone to adapt to its environment most likely involve strain-mediated fluid flow through the canalicular channels. Fluids can only be moved through bones by cyclic loading. The electrical signals are generated in two ways: by piezoelectricity and by streaming potentials. Streaming potentials derive from the bone fluid flow, which is generated by bone material deformation and blood circulation and is proportional to the strain rate.

Evidence has shown that an increase in venous pressure results in an increase in the passage of fluid from capillary to bone matrix [34]. Increased extravascular perfusion could be a factor in increasing periosteal bone formation. This flux of fluid may increase streaming potentials in bone, acting as a signal to bone cells to increase bone formation. Experiments by Lanyon [35] showed that cyclic loading induces more bone adaptation than static loading. Turner [36] performed experiments on cyclic loading of bone and determined that the stimulus for bone remodeling is proportional to the applied strain rate magnitude. Strain rate magnitude can be directly deduced from strain magnitude and frequency of loading. These phenomena can also be explained by bone electricity. It can thus be seen that both piezoelectricity and streaming potentials have relations to strain.

As for electrical signals generated by piezoelectricity, active research in the area of tissues such as living bone and collagen has shown that these materials are piezoelectric and that the piezoelectric properties of bone can also enable bone tissues to generate electrical signals that are proportional to the strain rate [32]. All these factors predict that the magnitude of the adaptive response of bone to loading should be proportional to the strain rate. For lower loading frequencies within the physiological range, experimental evidence shows this to be true. The data also suggest that activities that involve higher loading rates are also effective for increasing bone formation, even if the duration of the activity is short.

5.2.3 Bone Mechanostat

As described in Wikipedia, the bone Mechanostat is a model describing bone growth and bone loss. It was promoted by Harold Frost and described extensively in the *Utah Paradigm of Skeletal Physiology* in the 1960s [37]. The bone Mechanostat is a refinement of the law described by Julius Wolff [5]. To explain the strain-dependent behavior of bone tissues, Frost [38] proposed a "mechanostat" hypothesis describing bone modeling and remodeling, which was updated twice [37,39]. In his hypothetical model, mechanobiologic negative feedback mechanisms would work under the control of a subject's mechanical usage. In doing so they would adjust skeletal architecture in a way that tended to prevent mechanical usage from causing structural failure of skeletal tissues and organs.

It was proposed that mechanically dedicated message traffic would dominate the effects of most nonmechanical agents. Most (not all) nonmechanical agents would have permissive roles in affecting skeletal architecture and health. They could optimize or impair mechanical usage effects but could not replace or duplicate them. The mechanism of this process is shown in Figure 5.1, where **MU** denotes the skeleton's usual mechanical usage. Most systemic (**S**) agents reach the skeleton from the blood. Local (**L**) agents include local molecular biological agents, related phenomena, and local innervations. *MFL* indicates a mechanical feedback loop; here, each one is for modeling (*MFLm*) and remodeling (*MFLr*).

The updated Mechanostat indicated that signals were dependent on strains. Aided by sense systems that detect and process the signals, threshold ranges of the strain-dependent signals (the **MESm** for modeling and the **MESr** for disuse-mode remodeling) help to switch the two whole-bone-strength

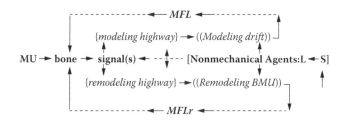

FIGURE 5.1
Mechanobiologic negative feedback mechanisms [35].

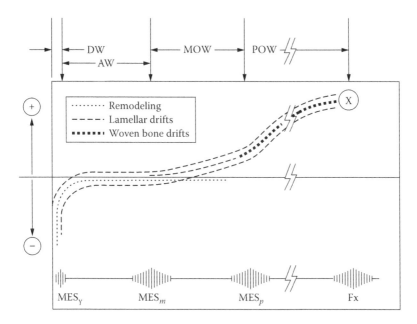

FIGURE 5.2
Combined modeling and remodeling effects on bone strength [37].

functions on and off. Figure 5.2 shows how these features would usually affect bone strength.

The horizontal line at the bottom suggests typical peak bone strains from zero on the left to the fracture strain on the right (**Fx**), plus the locations of the remodeling, modeling, and microdamage thresholds (**MESr, MESm,** and **MESp,** respectively). The horizontal axis represents no net gains or losses of bone strength. The lower dotted line curve suggests how disuse-mode remodeling would remove the bone next to marrow when strains remain below the **MESr** range; otherwise, the remodeling would tend to maintain the existing bone and its strength. The upper dashed line curve suggests how modeling drifts would begin to increase bone strength when strains enter or exceed the **MESm** range. The dashed outlines suggest the combined modeling and remodeling effects on bone strength. Beyond the **MESp** range, woven bone formation usually replaces lamellar bone formation. At the top, **DW** = disuse window and **AW** = adapted window, as in normally adapted young adults; **MOW** = mild overload window, as in healthy growing mammals; and **POW** = pathologic overload window. The strain span between the **MESr** and **MESm** represents the span between those features in the general biomechanical relations of bone.

It should be mentioned that the early work of Rubin and Lanyon [40] demonstrated increased activation of remodeling on endocortical and intracortical envelopes when strains were below a remodeling MES of 1,000 microstrains (με). Above this threshold, remodeling was inhibited and bone formation was initiated at the periosteal surface, which is consistent with the Mechanostat theory. Detailed experiments by Jee, Li, and Ke [41] indicated that a contralateral overloaded limb showed increased cancellous bone mass due to inhibited remodeling and increased bone formation rate. An increase of cortical bone mass on the periosteum was also observed. How, then, do bone tissues sense strain-related signals, and how are the thresholds distinguished? The foregoing theories do not explain these functions.

It is well known that two kinds of bone cells are involved in the bone remodeling (modeling) process: osteoblasts and osteoclasts. The former are responsible for bone formation and the latter for bone resorption. Evidence has shown that age-related bone osteoporosis differs between males and females. Accelerated bone loss in women during menopause is associated with reduced levels of estrogen. Estrogen receptors are abundantly expressed in osteocytes [42], but their expression is less in other cells of the osteoblast lineage, which suggests that osteocytes are likely involved in the regulation of estrogen-mediated bone remodeling (modeling). Osteocytes can then be considered as mechanosensitive bone cells [43,44] that are thought to activate other bone cells to initiate remodeling (modeling) activity in response to environmental stimuli, thus playing a key role in the regulation of bone remodeling (modeling) [45].

How the osteocytes participate in the remodeling (modeling) is still unknown. Investigations [27,28,46] showed that PDGF, IGF, BMP, and TGF-β exist in considerable quantities in bone matrix and play an important role in bone formation and remodeling (modeling). Normally, they are retained in osteocytes. Once the osteocytes are resorbed, the growth factors are released into the bone fluid and stimulate osteoblasts to refill the resorption cavities. Experiments have shown that a pulsed extremely low-frequency electromagnetic field can stimulate the multiplication of growth factors [29–31]. Thus, it can be seen that an electromagnetic field can influence the bone remodeling (modeling) process indirectly.

5.2.4 Adaptive Bone Modeling and Remodeling

As we can see from the Mechanostat model described earlier, it is a relatively mature hypothesis for a bone modeling and remodeling mechanism. But it is far from perfect. It does not describe how local mechanical signals are detected and how they are translated to bone formation and resorption. Nor does it indicate what the signals and nonmechanical agents are during the modeling and remodeling processes. Furthermore, although it distinguishes the strain thresholds of each mode, the reason for the existence of these thresholds is beyond its explanation capacity. To bypass these issues, electrical signals are defined as stimuli and growth factors as nonmechanical agents in this subsection. Then the modeling and remodeling processes of bone under electromagnetic loads are shown.

Compact bone structures are susceptible to failure when subjected to cyclic loadings, which often generate microfractures. So the remodeling process should have a function of repairing the damage in osteonal bone. It is known that the bone resorption function is mainly attributed to osteoclasts and bone formation to osteoblasts. When bone tissue is damaged, osteoclasts remove necrotic osteocytes. Growth factors such as BMP or TGF-β exist in latent forms in osteocytes. They are activated at the site of bone resorption by osteoclasts and released into the bone fluid. Osteoblasts are then stimulated by these growth factors to form bone and fill up resorption cavities.

It has been proposed that under normal circumstances the generation of damage by loading and its repair by remodeling can reach an equilibrium state in which the damage burden waiting to be repaired is tolerable [37]. If the loading increases, more microcracks are generated and more osteocytes are removed. This results in more growth factors in the bone fluid to accelerate bone formulation and maintain the equilibrium state. It has also been observed that when loadings are excessive, accelerated remodeling not only removes damage at a higher rate, but also increases the rate of damage production [47]. In contrast, a decrease in loading can also result in fewer microcracks and consequently lower presence of growth factors. It can thus be seen that bone tissue can remodel itself well to protect itself from damage and keep its mass unchanged.

But where do the two strain thresholds come from and why can the bone tissue change its mass and structure? Qu et al. [2] hypothesized that when the electrical signals change within a certain range, the quantities of growth factors hidden in osteocytes remain unchanged. When the electrical signals exceed this range, the quantities of growth factors in osteocytes will increase or decrease. Then the bone tissue begins to model itself. If the growth factors increase, more new bone tissue can be deposited and the bone mass will also increase. This can be considered as **MESm**. Similarly, **MESr** comes from the decrease of growth factors. The theory developed based on this hypothesis could be used to analyze the magnetoelectromechanical behavior of bone tissues in the modeling and remodeling processes.

5.3 A Mechanistic Model for Internal Bone Remodeling

In this section, the constitutive model presented in Hazelwood et al. [1] for bone remodeling, which includes a number of relevant mechanical and biological processes, is described. The model can be used to identify differences in modeling behavior as a volume element of bone is placed in disuse or overload.

5.3.1 Relationship between Elastic Modulus and Bone Porosity

This model was developed based on the relationship between Young's modulus **E** and porosity p obtained by fitting a polynomial to the results of Curry [48] and Rho, Ashman, and Turner [49]. By assuming a linear relationship between apparent density (ρ) and porosity (p), Hazelwood et al. [1] obtained a relationship between modulus and bone porosity as

$$E = (8.83 \times 10^5)p^6 - (2.99 \times 10^6)p^5 + (3.99 \times 10^6)p^4$$

$$- (2.64 \times 10^6)p^3 + (9.08 \times 10^5)p^2 - (1.68 \times 10^5)p + 2.37 \times 10^4 \tag{5.1}$$

in which E is in the unit of megapascals.

5.3.2 Porosity Changes

Based on the work of Martin [50], Hazelwood et al. [1] assumed that the rate of change of porosity ($\dot{p}$) is a function of the mean bone resorbing (Q_R) and refilling (Q_F) rates for each BMU, and the density of resorbing (N_R) and refilling (N_F) (the unit is BMUs/area)

$$\dot{p} = Q_R N_R - Q_F N_F \tag{5.2}$$

where the resorption ($Q_R = A/T_R$) and refilling ($Q_F = A/T_F$) rates are assumed to be linear in time, with A representing the area of bone resorbed by each BMU. T_R and T_F are the resorption and refilling periods, respectively.

In a cortical bone, the BMU forms a cylindrical canal about 2000 µm long and 150–200 µm wide. It gradually burrows through the bone at a speed of 20–40 µm/day. At the tip, on the order of 10 osteoclasts dig a circular tunnel (cutting cone) in the dominant loading direction. An activated osteoclast is able to resorb 200,000 µm³/day [50]. Then, several thousand osteoblasts will fill the tunnel (closing cone) to produce a (secondary) osteon of renewed bone. In this way, between 2% and 5% of cortical bone is remodeled each year [51]. In a trabecular bone, BMUs resorb and refill trenches rather than tunnels, but the process is similar and A is assumed to be the same for both cortical and trabecular bones.

The populations NR and NF were found by integrating over the appropriate resorption (T_R), reversal (T_I), and refilling (T_F) time intervals of the BMU activation frequency (f_a) history. For a histologic section, T_R is defined as the period from the moment the osteoclasts of a resorbing BMU enter the section to the time that resorption by these cells ceases in the section. The reversal or inactive period, T_I, follows and is the transition from osteoclastic to osteoblastic activity. Osteoblasts of the BMU then form new bone in the section during the refilling period, T_F. N_R is found by integrating f_a from time $t - T_R$ to t, where t is the present time

$$N_R = \int_{t-T_R}^{t} f_a(t')dt' \tag{5.3}$$

and N_F is found by integrating over the refilling period

$$N_F = \int_{t-(T_R+T_I+T_F)}^{t-(T_R+T_I)} f_a(t')dt' \tag{5.4}$$

Normal values of $T_R = 24$ days, $T_I = 8$ days, and $T_F = 64$ days were calculated from several histomorphometric studies [1,52,53].

5.3.3 BMU Activation Frequency

In Hazelwood et al. [1], the BMU activation frequency, f_a (BMUs/area/time), was assumed to be a function of disuse as well as of the existing state of damage. Also, because BMUs must start on a bone surface, f_a was taken to be a function of the internal surface area of the bone region. Specific surface area (internal surface area per unit volume, S_A) was determined from porosity using an empirical relationship [1], normalized to values between 0 and 1. Allowance was thus made for the greater potential for remodeling offered by large surface areas within a bone by letting

$$f_a = (f_{a(disuse)} + f_{a(disuse)})S_A \tag{5.5}$$

where $f_{a(disuse)}$ and $f_{a(damage)}$ represent contributions to f_a from disuse and damage, respectively.

5.3.4 Rate of Fatigue Damage Accretion

In their study, Hazelwood et al. defined the damage (D) as total crack length per section area of a bone. With this definition, they presented the fatigue damage accretion rate as

$$\dot{D} = \dot{D}_F - \dot{D}_R \tag{5.6}$$

where $\dot{D}_F$ and $\dot{D}_R$ represent the fatigue damage formation and removal rates, respectively. $\dot{D}_F$ is assumed to be proportional to the product of the strain range raised to a power and the loading rate (R_L, cycles per unit time) summed over n discrete loading conditions [47]:

$$\dot{D}_F = k_D \sum_{i=1}^{n} \varepsilon_i^q R_{Li} = k_D \Phi \tag{5.7}$$

where Φ is defined as the mechanical stimulus and k_D is a damage rate coefficient. For the sake of simplicity, Hazelwood et al. assumed that the strain, ε, was the principal compressive strain and that it returns to zero at the end of each load cycle, so that the strain range and peak strain are synonymous. The value for the exponent q was set at a nominal value of four based on the results of Whalen and Carter [54].

Then, Hazelwood et al. assumed that when BMUs and damage are randomly distributed in the bone, the damage removal rate is Df_aA. However, it was assumed that damage initiates BMU activation [55,56], so the efficiency of damage removal is greater than that of random remodeling. To allow for this, a damage removal specificity factor, F_s, is included in the equation for the damage removal rate

$$\dot{D}_R = Df_aAF_s \qquad (5.8)$$

where F_s was set to five based on the frequency with which microcracks were associated with new resorption cavities in the experiments of Mori and Burr [56]. In a state of equilibrium, $\dot{D}_F = \dot{D}_R$. This provides an estimate of the damage rate coefficient k_D as

$$k_D = D_0 f_{a0}AF_s/\Phi_0 \qquad (5.9)$$

where the subscript 0 indicates the initial equilibrium values assigned at the start of the simulation.

Using an average crack length of 0.088 mm [57] and the average crack density for a 40-year-old person [58], Hazelwood et al. obtained D_0 as 0.0366 mm/ mm^2. They also obtained an initial activation frequency, $f_{a0} = 0.00670$ BMU/ mm^2/day, from averaging several studies of cortical bone [52,53,59]. The initial mechanical stimulus (Φ_0) was estimated from cyclic strain levels necessary to "maintain" cortical bone mass in equilibrium [60,61]. A person walking 4.5 km/day with a modest 1.5 m stride (i.e., two steps) experiences about 3,000 cycles per day (cpd) (or roughly 1 million cycles per year) of lower extremity loading. Assuming $R_L = 3,000$ cpd to be typical, the results presented by Beaupre, Orr, and Carter [10] indicate that an equivalent cyclic strain of approximately 500 $\mu\varepsilon$ would constitute an equilibrium condition for cortical bones, with an initial mechanical stimulus of $\Phi_0 = 1.875 \times 10^{-10}$ cpd. Substituting these values into Equation (5.9) yielded $k_D = 1.85 \times 10^5$ mm/mm^2.

5.3.5 Disuse

To formulate the BMU response to disuse, Hazelwood et al. [1] adopted the "daily stress stimulus" approach of Carter, Fyhrie, and Whalen [62]. In their work, disuse was defined as stimulus values Φ below the equilibrium stimulus Φ_0. Thus, they used Φ both to calculate the damage formation rate ($\dot{D}_F = k_D\Phi$) and to quantify disuse as $\Phi_0 - \Phi$. Also, in disuse ($\Phi_0 > \Phi$), the refilling rate in Equation (5.2) was determined by setting the area of bone formed to $A[0.5 + 0.5\ (\Phi/\Phi_0)]$ to account for reduced refilling on bone surfaces in disuse states.

5.3.6 BMU Activation Frequency Response to Disuse and Damage

The relationships between the BMU activation frequency (f_a) and both disuse and damage were assumed to be sigmoidal, similar to responses found in pharmacological applications [1]. The coefficients in these functions were selected to fit the curves within known experimental data ranges. For disuse, Hazelwood et al. assumed that

$$f_{a(disuse)} = \frac{f_{a(max)}}{1 + e^{k_B(\Phi - k_C)}} \qquad \text{for } \Phi < \Phi_0 \qquad (5.10)$$

where $f_{a(max)}$ is the maximum activation frequency and k_B ($= 6.5 \times 10^{10}$cpd^{-1}) and k_C ($= 9.4 \times 10^{-11}$cpd $= \Phi_0/2$) are coefficients that define the slope and the inflection point of the curve, respectively.

The relationship between the BMU activation frequency and damage (D) was given [1] as

$$f_{a(damage)} = \frac{f_{a0}f_{a(max)}}{f_{a0} - (f_{a0} - f_{a(max)})e^{k_R f_{a(max)}(D-D_0)/D_0}} \qquad (5.11)$$

where D_0 is the initial equilibrium damage and $k_R = -1.6$ defines the shape of the curve. The nominal value for the maximum activation frequency, $f_{a(max)} = 0.50$ BMU/mm^2/day, used in Hazelwood et al. [1] was intentionally chosen to be higher than the highest average activation frequency (0.14 BMU/mm^2/day).

5.4 A Model for Electromagnetic Bone Remodeling

Based on the hypothesis described in Section 5.3, Qu et al. [2] developed a theoretical model for calculating the number of osteoclasts, N_R, and evaluating the corresponding remodeling behavior under electromagnetic loading. This section provides a description of the developments presented by these authors.

5.4.1 A Constitutive Model

First, N_R in Equation (5.3) is redefined by adding an additional term N_R^0 as:

$$N_R = \int_0^t f_a(t')dt' + N_R^0 \qquad (5.12)$$

where t is the time at which N_R is calculated and N_R^0 represents the number of BMUs that are required to resorb the naturally timeworn osteocytes and those that have been destroyed by microdamage. As proposed [2], the resorption of osteocytes is activated by the microdamage. So the BMU activation frequency, f_a (BMUs/volume/time), is assumed to be a function of the existing state of damage

$$f_a = f_{a(max)}(1 - e^{k_R\Phi}) \tag{5.13}$$

where $f_{a(max)} = 0.8$ BMU/mm^3/day [2], and $k_R = -1.6$ defines the shape of the curve. Φ is defined here as an environmental stimulus:

$$\Phi = C_{ij}\varepsilon_{ij}^q R_L + (C_i E_i + G_i B_i) f_e \tag{5.14}$$

Here, C_{ij}, C_i, and G_i are the damage rate coefficients. ε_{ij}, E_i, and H_i are strains, electrical field, and magnetic field, respectively. The value for the exponent q is set at a nominal 2/3. The mechanical loading rate, R_L, is assumed to be 3,000 cpd, and f_e is the frequency of the electromagnetic field.

The population N_F in Equation (5.2) is found by multiplying the quantities of resorbed osteocyte by k_F:

$$N_F = k_F N_R \tag{5.15}$$

where k_F is the correlation coefficient of the refilling BMUs, indicating the relation between the refilling and the resorbing process. k_F is defined as a piecewise function of Φ and p [2]:

$$k_F = \begin{cases} c_0 & \Phi_L \leq \Phi \leq \Phi_U \\ c_1 & \Phi < \Phi_U \\ (c_0 - c_2)\left(\dfrac{p}{p_0}\right)^n + c_2 & \Phi_L < \Phi \end{cases} \tag{5.16}$$

Considering that the quantity of growth factors retained in osteocytes changes along with the environmental loads, as mentioned before, Φ_L and Φ_U can be considered as **MESr** and **MESm,** respectively. When $\Phi_L \leq \Phi \leq \Phi_U$, the growth factors remain unchanged ($k_F = c_0$) and the bone tissue is in the remodeling state. When $\Phi < \Phi_U$, more growth factors ($k_F = c_1 > c_0$) are generated, which results in bone modeling. When $\Phi_L < \Phi$ fewer growth factors ($k_F = (c_1 > c_0)(p/p_0)^n + c_2 < c_0$) result in a disuse mode of bone tissue.

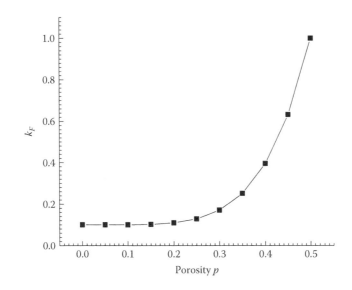

FIGURE 5.3
The relationship between k_F and p in disuse-mode remodeling.

The formula $(c_1 > c_0)(p/p_0)^n$ indicates the influence of biological factors. If this formula vanishes, when Φ converges to zero, p will approach 1, which means that all the bone tissue is resorbed. However, as is well known, although a mass of bone loss may be observed, bone tissue is not completely resorbed in the body of a patient who stays in bed for a long period of time. It is reasonable to predict that there must be some other factors contributing to bone remodeling as well as the mechanical factor. Qu et al. [2] assume it to be biological factors that prevent complete bone resorption. The porosity of the remaining bone tissue is assumed to be $p_0 = 50\%$ and $p_0 \leq p < 1$. The shape of the curve is defined with $n = 5$ (see Figure 5.3).

It can be seen from Figure 5.3 that as the porosity p increases, k_F also increases, which indicates that the bone tissue secretes more growth factors to deposit more bone material and restrain bone resorption.

This constitutive model is based on first-order, nonhomogeneous, nonlinear differential equations (5.2), which govern the evolutionary state variables of porosity and damage. The environmental stimulus Φ is regarded as the forcing function. The rate equation (5.2) involves implicitly the BMU activation frequency f_a, which itself is not an independent state descriptor, as it is algebraically related to p and Φ in Equations (5.13) and (5.16). The algorithm is implemented using a simple forward Euler scheme to integrate Equation (5.2) with respect to time. The integral in Equation (5.12) is calculated using the history of the daily average activation frequency. Then it can be used to analyze the bone modeling and remodeling process. Numerical simulation follows in the next section.

5.4.2 Numerical Examples

In the following analysis, Qu et al. [2] considered a cubic bone section subjected to uniaxial compressive pressure P and pulsed electromagnetic loads.
The side length a of the cubic section was assumed to be 1 cm. They also
assumed that the strain ε_{ij}, the electric field E_i, and the magnetic field H_i
all return to zero at the end of each load cycle, so their ranges and peaks
are the same. The model is given an initial porosity of 4.43% because this
allows equilibrium between the Haversian canals removed and added by
new BMUs [38,63]. This porosity produces an initial modulus of 17.8 GPa as
determined by Equation (5.1). A time increment of 0.5 day is examined to
integrate Equation (5.2). The state variables and constants used in Qu et al.
are shown in Table 5.1.

Bone resorption and formation can reach a proper equilibrium during the
remodeling process, which keeps the bone mass unchanged. It should be
mentioned that environmental factors affect the bone remodeling process.

TABLE 5.1

Model State Variables and Constants

State Variables		
E	Elastic modular (MPa)	
p	Porosity	
N_R	Number of resorbing BMUs (BMUs/mm³)	
N_F	Number of refilling BMUs (BMUs/mm³)	
f_a	BMU activation frequency (BMUs/mm³/day)	
ε_{ij}	Strain (με)	
Φ	Environmental stimulus (cpd)	
k_F	Correlation coefficient of the refilling and resorbing process	
Constants	**Description**	**Values used in Qu et al.**
V	Volume of bone tissues (mm³)	1,000
Q_R	Mean resorbing rate for each BMU (mm³/day)	2.0×10^{-4}
Q_F	Mean refilling rate for each BMU (mm³/day)	1.0×10^{-6}
$f_{a(max)}$	Maximum BMU activation frequency (BMUs/mm³/day)	0.8
k_r	Activation frequency dose–response coefficient	−1.6
R_L	Mechanical loading rate (cpd)	3,000
C_{ij}, C_i, G_i	Damage rate coefficients	0.04, 0.08, 0.04
f_e	Frequency of electromagnetic field (Hz)	100
N_R^0	Number of naturally timeworn osteocytes (BMUs/mm³)	0.4
c_0	Value of k_F during remodeling process	1.0
c_1	Value of k_F during modeling process	1.2
c_2	Value of k_F in disuse-mode remodeling	0.1
p_0	Porosity of unresorbable bone tissues	0.5

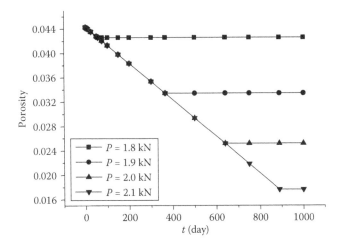

FIGURE 5.4
Variation of porosity *p* for several overloads.

An increase in Φ can result in faster bone remodeling, and vice versa. Qu et al. [2] investigated only the modeling and disuse-mode remodeling of bone tissues. They distinguished the following six loading cases:

- Case 1: $P = 1.8, 1.9, 2.0, 2.1$ kN, $E_i = 0$, $H_i = 0$. The results for this loading case are shown in Figure 5.4. It can be seen that overloads can activate bone modeling. The porosity of bone tissue decreases when the environmental stimuli exceed the **MESm,** which is defined as the modeling threshold. Overloads result in a denser and stronger bone structure. The elastic module *E* increases due to the decrease in porosity. Then the environmental stimulus decreases at the same time as the strains become smaller. When porosity returns to the remodeling threshold it will not change any further. The result shows that bone tissue can model itself to force its strains to revert to the remodeling range. It can also be seen that the greater the pressure is, the less porous the bone material will be. But it should be mentioned that if the loading is so great that the strain cannot be reduced to the remodeling range when the porosity reaches its lower limit, the bone structure will model itself in another way.
- Case 2: $P = 0, 0.05, 0.10, 0.15$ kN, $E_i = 0$, $H_i = 0$. This case is investigated to demonstrate the bone disuse-mode remodeling process. The corresponding results are presented in Figure 5.5. It can be seen from the figure that, as the loadings decrease, the bone materials become more porous to resist the decrease of the environmental stimulus. But, as already mentioned, the porosity of a bone tissue should be

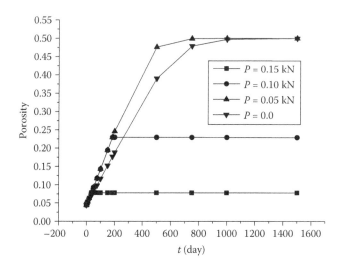

FIGURE 5.5
Variation of porosity *p* in the disuse-mode case.

below a certain (or critical) level. Here we define the value as 50%. When the porosity approaches this value, biological factors will stimulate the osteocytes to excrete more growth factors to resist the loss of bone mass. This is clearly shown in Figure 5.5. It can also be shown that when the loading vanishes, the velocity of bone remodeling is not as fast as in bone materials subjected to compressive loads. This can be attributed to the lack of environmental stimuli, resulting in a reduction of osteoclasts. Then fewer osteocytes are resorbed and fewer growth factors are released, slowing down the loss of bone mass.

- Case 3: $P = 1.0$ kN, $E_i = 1, 10, 50, 100$ V/m, $f_e = 15$ Hz. Figure 5.6 shows the effect of electrical loading on the bone modeling process. It can be seen that when environmental stimuli are insufficient, the remodeling state of bone tissue will remain unchanged. As the electrical loading increases to a particular level, bone modeling can be triggered. A more intense electrical field can produce a less porous and denser bone structure. But when the electrical loading is sufficiently high, a further increase will have very little effect on the bone modeling process. This is also due to an insufficiency of osteoclasts. The capacity of the body to produce osteoclasts restricts the upper limit of growth factors. So the electrical loading that can effectively stimulate bone modeling must have both an upper and a lower limit. However, all these conclusions are based on the hypothetical model. At this stage, we cannot give the exact values of these thresholds, which require further experimental investigation in this field.

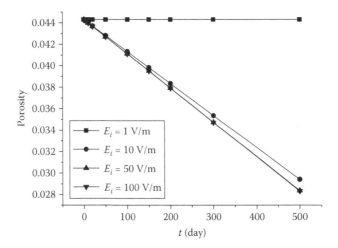

FIGURE 5.6
Variation of porosity p for several electrical changes.

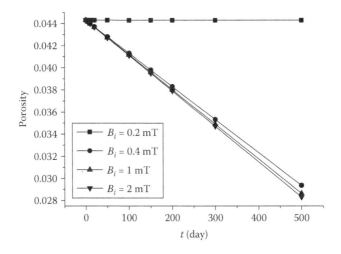

FIGURE 5.7
Variation of porosity p for several magnetic loadings.

On the other hand, the result also indicates that bone materials become increasingly dense after electrical fields are loaded. Although a remodeling balance may be finally reached, bone tissues exposed to an electromagnetic field for a long time may incur a high risk of bone hypertrophy.

- Case 4: $P = 1.0$ kN, $B_i = 0.2, 0.4, 1, 2$ mT, $f_e = 15$ Hz. Figure 5.7 shows the effect of magnetic loading on the bone modeling process. The results are similar to those for electrical loading. There are upper and lower

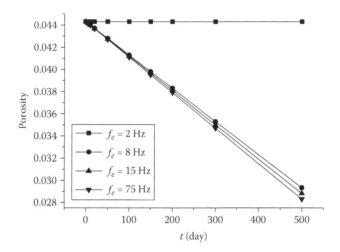

FIGURE 5.8
Variation of porosity p for bone modeling in a bone subjected to electromagnetic fields of different frequencies.

thresholds in magnetic loads, and long exposure to a magnetic field can also cause bone hypertrophy.

- Case 5: $P = 1.0$ kN, $E_i = 10$ V/m, $B_i = 2$ mT, $f_e = 2, 8, 15, 75$ Hz. This case concerns the frequency of electromagnetic fields loaded on bone tissues. The results (see Figure 5.8) are similar to the previous two cases. All three figures (Figures 5.6–5.8) indicate that an electromagnetic field can trigger bone modeling. The effect of the loading on bone modeling is dependent on the intensity and frequency. This feature can be applied in clinical practice to treat bone diseases, such as osteoporosis and nonunion, with a pulsed, extremely low-frequency electromagnetic field. As the results in this case show, an electromagnetic field can influence bone modeling. This effect has been used in the treatment of bone disease. As yet we do not know how a bone tissue remodels itself after electromagnetic loads cease. However, post-treatment maintenance is just as important as curative effects. Here we consider three cases to study the remodeling and disuse mode, respectively. $P = 0.15$ kN defines the disuse mode and the two later cases are remodeling examples. Electromagnetic fields are loaded and are unloaded after 500 days.

- Case 6: $P = 0.15, 0.2, 1.2$ kN, $E_i = 10$ V/m, $B_i = 2$ mT, $f_e = 15$ Hz. Figure 5.9 shows the simulation results. As we can see from the figure, after the electromagnetic field is loaded, the porosity of bone tissue decreases due to bone modeling, but unloading the electromagnetic field results in different effects. In disuse mode, the bone structure becomes more porous and finally returns to its initial state. In remodeling mode,

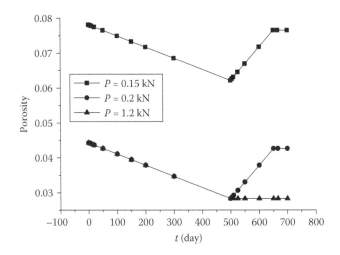

FIGURE 5.9
Variation of porosity *p* for bone material subjected to several multifield loads.

when the loading is relatively small, the variation in porosity is similar to that in disuse mode. On the other hand, if the loading is large enough, the porosity will remain unchanged. The first result is due to an insufficiency of environmental stimuli. After the electromagnetic field is removed, there is no other loading to stimulate bone modeling except the initial mechanical loading. Thus, bone tissue reverts to disuse-mode remodeling and bone mass loss is triggered again. This leads to the conclusion that, although electromagnetic treatment is effective, active exercise is necessary to maintain the curative effect. The second result can be explained as follows. The electromagnetic load makes the bone structure more rigid. The initial mechanical loading cannot stimulate bone remodeling sufficiently after the electromagnetic field is unloaded. The bone tissue begins to remodel itself in disuse mode, which causes bone loss and increased porosity. This indicates that although an electromagnetic field can induce bone hypertrophy, it can be automatically cured after the field is removed. But this occurs only in some cases. In other cases, as shown in the third case ($p = 1.2$ kN), the bone mass gain is permanent.

5.5 Bone Surface Modeling Model Considering Growth Factors

Section 5.4 described a model dealing with internal bone remodeling due to electromagnetic loading. Extension to the case of surface modeling was reported in He et al. [3] and is reviewed in this section.

5.5.1 Equations of Growth and Remodeling

Based on the hypothesis presented in Section 5.3, He et al. [3] proposed a computable model for the process of surface bone remodeling. In that model they assumed that N_R consists of following two parts:

$$N_R = N_R^1 + N_R^0 \tag{5.17}$$

where N_R^1 is the population of environmentally stimulated BMUs that is assumed to be a function of the existing state of damage:

$$N_R^1 = N_{R(\max)}^1 (1 - e^{k_r \Phi}) \tag{5.18}$$

where $N_{R(\max)}^1$ is the maximum number of environmentally stimulated BMUs, $N_{R(\max)}^1 = 0.8$ BMU/mm^2, $k_r = -1.6$, and the environmental stimulus Φ is defined by Equation (5.14).

For convenience, N_F in Equation (5.2) is divided into two major parts: N_F^i and N_F^s, where N_F^i represents the number of refilling BMUs in internal bone remodeling and N_F^s denotes the number of refilling BMUs in surface bone remodeling. Then, Equation (5.2) is rewritten as

$$\dot{p} = Q_R N_R - Q_F N_F^i \tag{5.19}$$

where N_F^i is determined by the following [64]:

$$N_F^i = \begin{cases} N_F & \text{if } N_F \leq N_0 \\ N_0 & \text{if } N_F > N_0 \end{cases} \tag{5.20}$$

As in Equation (5.15), the total number of refilling BMUs N_F^s can be found by multiplying the quantity of resorbed osteocytes by k_f:

$$N_F^s = k_f N_R \tag{5.21}$$

where k_f is defined in Equation (5.15). When bone tissues are overloaded, $k_f = c_1$.

The formulation is for the internal bone remodeling discussed in Section 5.4. As noted, when the porosity of bone structure is reduced to a certain magnitude, the growth factors exceed the quantities consumed by the internal bone remodeling and the excess is transported to the surface of the bone structure, where new bone material is deposited. He et al. [3] defined this threshold as N_0, which is dependent on porosity p. Then they assumed a linear relation between the porosity p and the population N_0:

$$N_0 = k_s p \tag{5.22}$$

where k_s is the proportional coefficient that denotes the maximum growth factors consumed by the internal bone remodeling per unit volume. Because the variation of bone surfaces is very small, the surface bone remodeling rate can be written as

$$-\frac{da}{dt} = k_e \left(N_F - N_0 \right), \qquad \frac{db}{dt} = k_p \left(N_F - N_0 \right) \tag{5.23}$$

where k_e is the coefficient of the endosteal remodeling rate and k_p is the coefficient of the periosteal remodeling rate.

Consider a hollow circular cylinder of bone that is subjected to a quasistatic axial load P, an external pressure P_t, an electric load φ_a (and/or φ_b), and a magnetic load ψ_a (and/or ψ_b). The related equations for constitutive relation, equilibrium equation, boundary conditions, and expressions of strains defined by Equations (3.71), (3.74)–(3.76), (3.6), (3.77), and (3.82)–(3.86) are still applicable here.

To show the results more clearly, changes of the inner and outer surfaces are again written in terms of the following nondimensional parameters (see Equation 4.17):

$$\varepsilon = \frac{a}{a_0} - 1, \quad \eta = \frac{b}{b_0} - 1 \tag{5.24}$$

It is obvious that Equations (3.82)–(3.86) show that all the variables are dependent on r only. Therefore, the environmental stimulus must also be a function of r. As the bone remodels itself, the stimuli of the inner and the outer surfaces vary in the radial direction, which results in variation of the BMU activation frequencies and the quantity of growth factors in different locations of the bone. However, He et al. [3] assumed that the growth factors can be uniformly distributed to the bone structures by means of the fluid flow. The homogenized quantities of the growth factors can be shown as

$$N_R^1 = \frac{\displaystyle\int_0^{2\pi} \int_{R_1}^{R_2} N_{R(\max)}^1 \left(1 - e^{k_r \Phi} \right) r \, dr \, d\theta}{\pi \left(b^2 - a^2 \right)} \tag{5.25}$$

where R_1 denotes the inner radius at which the environmental stimulus exceeds the remodeling threshold and R_2 denotes the outer radius at which the environmental stimulus exceeds the remodeling threshold.

He et al. [3] next simulated the bone surface remodeling process numerically, based on the preceding model.

Once the strains, electrical field, and magnetic field are obtained, the differential equations (5.2) and (5.23) can be rewritten as

$$\frac{dp}{dt} = f_1(t, p, a, b),$$

$$\frac{da}{dt} = f_2(t, p, a, b), \tag{5.26}$$

$$\frac{db}{dt} = f_3(t, p, a, b)$$

where

$$f_1(t, p, a, b) = Q_R N_R - Q_F N_F^i \tag{5.27}$$

$$f_2(t, p, a, b) = -k_e(N_F - N_0) \tag{5.28}$$

$$f_3(t, p, a, b) = -k_p(N_F - N_0) \tag{5.29}$$

The following fourth-order Runge-Kutta integrating process can be used to obtain the numerical solution:

Step 1:

$$K_1^p = f_1(t, p^t, a^t, b^t)$$

$$K_1^a = f_2(t, p^t, a^t, b^t) \tag{5.30}$$

$$K_1^b = f_3(t, p^t, a^t, b^t)$$

Step 2:

$$K_2^p = f_1\left(t + \frac{\Delta t}{2}, p^t + \frac{\Delta t}{2} K_1^p, a^t + \frac{\Delta t}{2} K_1^a, b^t + \frac{\Delta t}{2} K_1^b\right)$$

$$K_2^a = f_2\left(t + \frac{\Delta t}{2}, p^t + \frac{\Delta t}{2} K_1^p, a^t + \frac{\Delta t}{2} K_1^a, b^t + \frac{\Delta t}{2} K_1^b\right)$$

$$K_2^b = f_3\left(t + \frac{\Delta t}{2}, p^t + \frac{\Delta t}{2} K_1^p, a^t + \frac{\Delta t}{2} K_1^a, b^t + \frac{\Delta t}{2} K_1^b\right)$$

Step 3:

$$K_3^p = f_1\left(t + \frac{\Delta t}{2}, p^t + \frac{\Delta t}{2}K_2^p, a^t + \frac{\Delta t}{2}K_2^a, b^t + \frac{\Delta t}{2}K_2^b\right)$$

$$K_3^a = f_2\left(t + \frac{\Delta t}{2}, p^t + \frac{\Delta t}{2}K_2^p, a^t + \frac{\Delta t}{2}K_2^a, b^t + \frac{\Delta t}{2}K_2^b\right)$$

$$K_3^b = f_3\left(t + \frac{\Delta t}{2}, p^t + \frac{\Delta t}{2}K_2^p, a^t + \frac{\Delta t}{2}K_2^a, b^t + \frac{\Delta t}{2}K_2^b\right)$$

Step 4:

$$K_4^p = f_1\left(t + \Delta t, p^t + \Delta t K_3^p, a^t + \frac{\Delta t}{2}K_3^a, b^t + \frac{\Delta t}{2}K_3^b\right)$$

$$K_4^a = f_2\left(t + \Delta t, p^t + \Delta t K_3^p, a^t + \frac{\Delta t}{2}K_3^a, b^t + \frac{\Delta t}{2}K_3^b\right)$$

$$K_4^b = f_3\left(t + \Delta t, p^t + \Delta t K_3^p, a^t + \frac{\Delta t}{2}K_3^a, b^t + \frac{\Delta t}{2}K_3^b\right)$$

Then, the variables at $t + \Delta t$ are calculated by

$$p^{t+\Delta t} = p^t + \frac{\Delta t}{6}\left(K_1^p + 2K_2^p + 2K_3^p + K_4^p\right)$$

$$a^{t+\Delta t} = a^t + \frac{\Delta t}{6}\left(K_1^a + 2K_2^a + 2K_3^a + K_4^a\right)$$

$$b^{t+\Delta t} = b^t + \frac{\Delta t}{6}\left(K_1^b + 2K_2^b + 2K_3^b + K_4^b\right)$$

The algorithm described here is used in the next section.

5.5.2 Bone Remodeling Simulation

For simplicity, He et al. [3] considered a section of hollow cylindrical bone that is subjected to axial compressive pressure P, transverse pressure p_t, and pulse electromagnetic loads. We also assume that the strain ε_{ij}, the electric field E_i, and the magnetic field H_i all return to zero at the end of each load cycle, so their ranges and peaks are the same. The model is given an initial porosity of 8%. The state variables and constants are shown in Table 5.2.

As a numerical illustration of the bone surface remodeling process, He et al. [3] considered a femur with $a = 25$ mm and $b = 35$ mm. The elastic moduli

TABLE 5.2

Model State Variables and Constants

	State variables	
N_R^1	Number of resorbing BMUs due to microdamage	
N_F^s	Number of refilling BMUs in surface bone remodeling	
N_F^i	Number of refilling BMUs in internal bone remodeling	
N_0	Maximum of refilling BMUs consumed by internal remodeling strains	
	Constant values used in He et al. [3]	
$N_{R(\max)}^1$	Maximum BMUs that can be generated by the body	0.8
k_f	Correlation coefficient of the refilling and resorbing process	1.8
k_s	Proportional coefficient	0.6
k_e	Inner surface bone remodeling rate coefficient	0.25
k_p	Outer surface bone remodeling rate coefficient	0.4
f_e	Frequency of the electromagnetic field	2
Φ_U	Upper threshold of bone remodeling	0.0048
N_R^0	Number of naturally timeworn osteocytes	0.4
c_1	Value of k_f during the remodeling process	3.6×10^2

were calculated as $c_{ij} = c_{ij}^0(1-p)^n$, $e_{ij} = e_{ij}^0(1-p)^n$, and $\alpha_{ij} = \alpha_{ij}^0(1-p)^n$, where c_{ij}^0, e_{ij}^0 and α_{ij}^0 are constants that are related to the properties of materials and $n = 3$ [48]. The reference values of material properties are assumed as follows:

$$c_{11}^0 = 15\,\text{GPa},\ c_{12}^0 = c_{13}^0 = 6.6\,\text{GPa},\ c_{33}^0 = 12\,\text{GPa},\ c_{44}^0 = 4.4\,\text{GPa},\ e_{15}^0 = 1.14\,\text{C/m}^2,$$
$$\alpha_{15}^0 = 550\,\text{N/Am}.$$

Bone is usually subjected to pressure, and greater pressure can strengthen it. It has been reported that osteoclasts migrate to the positive electrode in an electric field, whereas osteoblasts migrate to the negative electrode [65]. Therefore, all the remodeling rate coefficients are negative. They can be shown as

$$C_{rr} = -0.06\,\text{m/day},\ C_{\theta\theta} = -0.03\,\text{m/day},\ C_{zz} = -0.04\,\text{m/day}$$

$$C_{zr} = -0.01\,\text{m/day},\ C_r = 3 \times 10^{-7}\,\text{V}^{-1}\text{m}^2/\text{day},\ \text{and}\ G_r = 4 \times 10^{-8}\,\text{A}^{-1}\text{m}^2/\text{day}$$

The initial inner and outer radii are again assumed to be

$$a_0 = 25\,\text{mm},\ b_0 = 35\,\text{mm}.$$

Four loading cases are considered to examine the influence of axial pressure, transverse pressure, electrical field, and magnetic field on bone remodeling. The results are presented next.

5.5.2.1 Effect of Axial Pressure on Bone Remodeling Process

To illustrate the influence of axial pressure on bone remodeling, the variation rate of the radii of the inner surface and the changes in the porosity of the bone material are calculated and shown in Figures 5.10 and 5.11. The applied loadings are P = 1.6, 1.8, 2.0, 2.2, and 2.4 kN; no other loadings are applied.

Figures 5.10 and 5.11 show that axial pressure can strengthen a bone tissue by the deposition of new bone material in and on the surface of the bone. Figure 5.10 shows the influence of axial pressure on internal bone remodeling. The porosity of the bone tissue decreases when the environmental stimuli exceed the MESm, which is defined as the remodeling threshold. Overloads make the bone material denser and stronger. The elastic modulus also increases, resulting in a decrease of strain on the bone structure. The environmental stimulus decreases at the same time, as the strains become smaller.

When the environmental stimulus returns to the remodeling threshold, the porosity will not change any further. This shows that a bone tissue can model itself to force the environmental stimulus to revert to the remodeling range. However, when the axial loading is sufficiently high, a further increase has very little effect on the bone remodeling process. Two factors

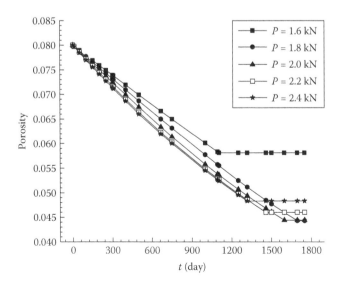

FIGURE 5.10
Variation of porosity p for several axial overloads.

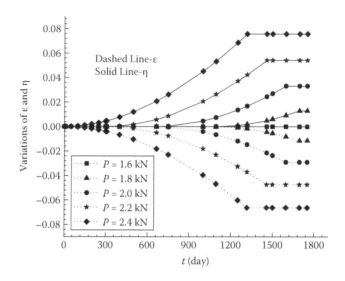

FIGURE 5.11
Variations of ε and η for several axial overloads.

are responsible for this. One is an insufficiency of osteoclasts. The capacity of the body to produce osteoclasts restricts the upper limit of growth factors. Therefore, the axial loading that can effectively stimulate bone remodeling must have an upper limit.

The other factor is saturation of growth factors, which seems to contradict the former factor. Nevertheless, as mentioned previously, when the porosity of a bone is sufficiently reduced, the quantity of growth factors will exceed the capacity of the bone to consume. The unconsumed growth factors are then transported to the surface by fluid flows. At that time, an increase in the environmental stimulus can accelerate surface bone remodeling only, not internal remodeling. Thus, the internal remodeling rate also has an upper threshold.

Figure 5.11 shows the influence of the axial pressure on surface bone remodeling. The results indicate that the inner surface of the bone decreases and the outer surface increases when the axial loadings exceed the MESm. The effect of axial pressure on the inner and outer surfaces indicates that axial pressure can increase the cross-sectional area of the bone and, consequently, a thicker and stronger bone structure can be obtained, which can decrease the strain on the bone structure. Furthermore, greater pressure can result in more change in the surface of the bone and accelerate the recovery of injured bone.

If we examine the two figures together, the results are interesting. In the case of internal bone remodeling, when axial pressure increases, the porosity initially increases and then becomes less as another equilibrium state is finally achieved. The remodeling rate (the slope of the line) does not change with time when P = 1.6 kN. In other cases, the remodeling rate decreases with the remodeling process. Nevertheless, the remodeling rate of surface remodeling increases during the remodeling process. The surface remodeling finally reaches an equilibrium state at the same time as the internal remodeling, but the time at which the surface remodeling process begins is different. The greater the axial loading is, the earlier the onset of surface remodeling will be.

These results can be illustrated as follows. We can see that surface bone remodeling is not triggered because of a comparative insufficiency of the growth factors that are generated by the body when P = 1.6 kN. Therefore, the internal bone remodeling rate remains unchanged. However, as the environmental stimulus increases, more growth factors are secreted and finally exceed the consumption capacity of the internal bone remodeling. Some of the growth factors are then transported to the bone surface, where new bone material is deposited and surface remodeling is triggered, resulting in a decrease of the rate of internal bone remodeling. As the porosity of bone structure decreases, more and more growth factors are transported to the bone surface. Hence, the internal bone remodeling rate decreases as the surface rate increases, and they finally cease at the same time when the remodeling process reaches equilibrium.

This is different from the following cases because of the inhomogeneity of the environmental stimulus. Moreover, if the environmental stimulus increases, more growth factors are released to the bone fluid. It is obvious that saturation can then more easily be reached and surface remodeling triggered earlier. Coupled remodeling is obviously more effective than separate remodeling. The final porosity is then increased as the environmental stimulus increases. Bone tissue seems to have the capacity to prevent the bone structure from becoming so dense that the bone fluid cannot be transported effectively. Determination of the exact threshold is still an open question; further theoretical and experimental investigation is needed.

5.5.2.2 *Effect of Transverse Pressure on Bone Remodeling Process*

In this section we consider the influence of transverse loads on the bone remodeling process. The results are shown in Figures 5.12 and 5.13. The applied loadings are P = 1.2 kN, p_t = 0.1, 0.2, 0.3, 0.4, 0.5 Mpa; no other loadings are applied.

It can be seen from Figures 5.12 and 5.13 that transverse pressures have some effects on the bone remodeling process that are similar to those of axial loadings. For example, transverse pressure can also make the bone structure stronger and denser. It can increase the cross-sectional area of a bone and make it thicker. Moreover, the change rate increases as the magnitude of the loading increases. For surface remodeling, a difference in trigger time also exists. However, as noted, other differences are evident. First, for internal

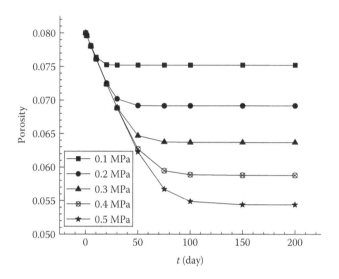

FIGURE 5.12
Variation of porosity p for several transverse pressures.

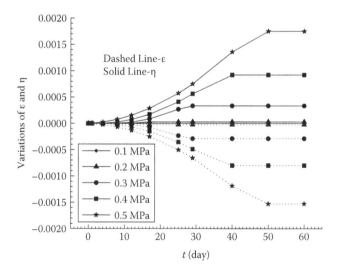

FIGURE 5.13
Variations of ε and η for several transverse pressures.

bone remodeling, a gradually convergent process can be observed compared to the former case; for surface remodeling, the rate does not increase from the beginning to the end of the remodeling process. It first increases and subsequently decreases. The time cost for internal bone remodeling also varies from that for surface remodeling. Surface remodeling costs less time than internal remodeling.

These differences result from the inhomogeneity of the environmental stimulus. As the transverse pressure is loaded, an inhomogeneous strain field is generated, which results in the inhomogeneous distribution of the environmental stimulus. The quantity of growth factors released to the bone fluid then varies in different areas of the bone structure. As the environmental stimulus decreases due to the bone remodeling process, the inhomogeneity disables the bone tissues in different areas of the bone structure from reaching the MESm at the same time. This means that the bone tissue in some areas of the bone cannot excrete excess growth factors (i.e., redundant growth factors after the refilling process) to deposit additional bone material because the environmental stimulus is within the remodeling range.

However, additional growth factors can still be generated in the remaining areas at the same time, and new bone material deposition will continue until all of the bone tissues return to the remodeling mode. The amount of bone tissue in the remodeling state will then diminish as the bone remodeling process goes on. Thus, the gradual divergence of internal bone remodeling and the rate change of surface remodeling can be easily understood. The reversal of the surface remodeling rate occurs when some of the bone tissue reverts to the remodeling mode. In the case of axial loading, the environmental stimulus is distributed homogeneously, which means that all of the bone tissue can revert to the remodeling mode at the same time, and internal and surface remodeling cease simultaneously. However, in this case, when surface remodeling ceases, not all of the bone tissue stops excreting additional growth factors. Internal bone remodeling will continue until all of the bone tissue is in the remodeling mode. Thus, the surface and internal remodeling cease at different times. It should be mentioned that, as we assumed before, growth factors can be homogeneously distributed in bone structure due to the flow of bone fluid, so a homogeneous bone structure is maintained.

5.5.2.3 Effect of an Electrical Field on Bone Remodeling Process

As earlier investigations have shown that the influences of electrical and magnetic fields are similar [2,66], we here consider only the electrical field. The results are shown in Figures 5.14 and 5.15. The applied loadings are $P = 1.2$ kN, $\varphi = 5, 10, 15, 20, 25$ V, $f_e = 2$ Hz, and no other loadings are applied; here, $\varphi = \varphi_b - \varphi_a$.

Figures 5.14 and 5.15 show the effect of electrical loading on the internal and surface bone remodeling process. It can be seen that as the electrical loading increases to a certain level, bone remodeling can be triggered. An electrical field can model the bone structure to fit its environment so that the loading does not break or harm it. Electrical fields can also make bone structure stronger and denser and increase the cross-sectional area of bone and make it thicker. The change rates increase as the intensity of the loading increases. All these features are effective methods to return a bone structure to the remodeling mode.

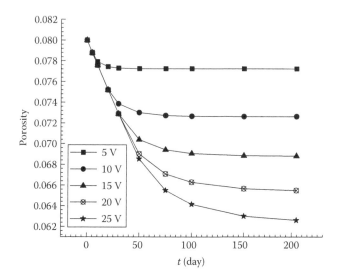

FIGURE 5.14
Variation of porosity p subjected to different electrical fields.

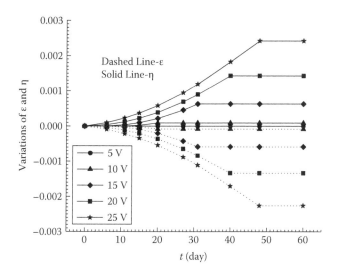

FIGURE 5.15
Variations of ε and η subjected to different electrical fields.

Furthermore, gradual divergence of the internal remodeling rate, differences in the trigger time, reversal of the change of the rate of surface remodeling, and different cessation times can also be seen in this case due to the inhomogeneous distribution of the environmental stimulus, as in the case of transverse pressure. However, an electrical field seems to affect bone remodeling more moderately than does transverse pressure. The changes in the remodeling rate under the electrical field are smoother and less marked.

5.5.2.4 Effect of Multifield Loadings on Bone Remodeling Process

The results in this section show bone remodeling behavior under multifield loadings. This case simulates a bone structure that is subjected to axial and transverse pressure and a pulsed magnetic field. Several magnetic loadings of different intensity are examined. The results can be seen in Figures 5.16 and 5.17. The applied loadings are $P = 1.5$ kN, $p_t = 0.2$ MPa, $\psi = -0.2, -0.1, 0,$ $0.1, 0.2$A, $f_e = 2$ Hz; here, $\varphi = \varphi_b - \varphi_a$.

Figures 5.16 and 5.17 show that a positive magnetic field can stimulate the bone remodeling process, whereas a negative magnetic field can do the opposite. When the bone structure is loaded by a positive magnetic field, both the density and cross-sectional area increase, and stronger and thicker bone is generated. The more intense the magnetic field is, the stronger and thicker the bone will be. However, a negative magnetic field can weaken bone remodeling, making bone thinner and more porous. A positive magnetic field can play the same role as an increase in transverse loading, whereas a negative magnetic field can replicate the effect of a decrease in such loading. The effect of magnetic loading is smooth and gentle. Used as a method to control remodeling, it is therefore an appropriate way of dealing with bone conditions.

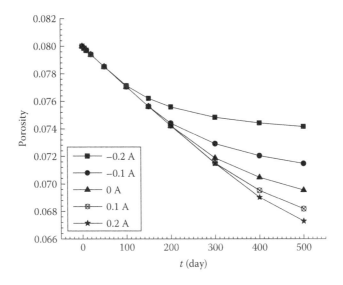

FIGURE 5.16
Variations of porosity p for several different multifield loadings.

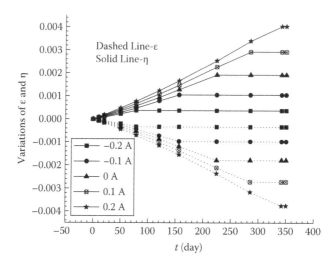

FIGURE 5.17
Variations of ε and η for several different multifield loadings.

5.6 Bone Remodeling Induced by a Medullary Pin

Consider again the problem of a hollow circular cylinder inserted by a medullary pin, as discussed in Section 3.5 (see Figure 5.18). The boundary conditions are slightly different from those employed in Section 3.5 and are given
as follows [64]:

$$\sigma_{rr} = -p(t), \quad \sigma_{r\theta} = \sigma_{rz} = 0, \quad \varphi = \varphi_a, \quad \psi = \psi_a \quad \text{at } r = a,$$
$$\sigma_{rr} = \sigma_{r\theta} = \sigma_{rz} = 0, \quad \varphi = \varphi_b, \quad \psi = \psi_b \quad \text{at } r = b \tag{5.31}$$

$$\int_S \sigma_{zz} dS = -P_2 \tag{5.32}$$

5.6.1 The Solution of Displacements and Contact Force *p(t)*

Following the procedure used in Section 3.5, we obtain the solution of
displacement u_2 as

$$u_2 = \frac{a}{F_3^*}\left(\frac{c_{33}p(t)b^2}{b^2 - a^2} + \frac{c_{13}P_2}{\pi(b^2 - a^2)}\right) + \frac{ab^2 p(t)}{(c_{11} - c_{12})(b^2 - a^2)} \tag{5.33}$$

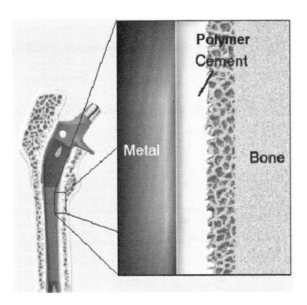

FIGURE 5.18
A cylindrical bone with a medullary pin inserted [64].

Assuming that the radius of the pin is $a_0 + \delta/2$, where a_0 is the radius of the cylindrical bone, we obtain Equation (3.63) and rewrite it here for the sake of convenience

$$a_0 + \frac{\delta}{2} + u_1 = a_0 + u_2 \tag{5.34}$$

where u_1 is given by the following [64]:

$$u_1 = \frac{a}{E}\left[\frac{vP_1}{\pi(b^2 - a^2)} - (1 - v)p(t)\right] \tag{5.35}$$

for an isotropic pin, where E and v are, respectively, Young's modulus and Poisson's ratio. Making use of Equations (5.33)–(5.35), we obtain the solution of contact force $p(t)$ as

$$p(t) = \frac{\dfrac{\delta}{a} + \dfrac{vP_1}{E\pi(b^2 - a^2)} - \dfrac{c_{13}}{F_3^*} \cdot \dfrac{P_2}{\pi(b^2 - a^2)}}{\left(\dfrac{c_{13}}{F_3^*} + \dfrac{1}{c_{11} - c_{12}}\right) \cdot \dfrac{b^2}{b^2 - a^2} + \dfrac{1 - v}{E}} \tag{5.36}$$

5.6.2 A Constitutive Remodeling Model

For internal bone remodeling, the change of porosity is still defined by

$$\dot{p} = U_R - U_F \tag{5.37}$$

where U_R and U_F are, respectively, the total resorbing rate and refilling rate. They are assumed to be linear functions of the density of resorbing (N_R^i, N_R^s) and (N_F^i, N_F^s) refilling as

$$U_R = k_R N_R^i,$$
$$U_F = k_F(N_R^i + N_R^s) \tag{5.38}$$

where

k_R and k_F are, respectively, the correlation coefficient of bone resorbing and refilling rates

k_F is defined by Equation (5.16)

N_R^i and N_R^s are the numbers of resorbing BMUs in internal and surface bone remodeling

N_F^i and N_F^s are the numbers of refilling BMUs in internal and surface bone remodeling, respectively

Similarly to Equation (5.25), N_R^i here is determined by

$$N_R^i = \frac{\int_0^{2\pi} \int_{R_1}^{R_2} N_{R(\max)}\left(1 - e^{k_R\Phi}\right) r\, dr\, d\theta}{\pi\left(b^2 - a^2\right)} \tag{5.39}$$

and N_R^s is assumed to be a linear function of the surface velocity U_e as

$$N_R^s = \frac{2\pi a U_e}{k_R \pi \left(b^2 - a^2\right)} = \frac{2 a U_e}{k_R \left(b^2 - a^2\right)} \tag{5.40}$$

where U_e is calculated by

$$U_e = k_R^s\left[1 - e^{k_R(\Phi - \Phi_L)}\right] \tag{5.41}$$

or

$$-\frac{da}{dt} = k_R^s\left[1 - e^{k_R(\Phi_L - \Phi)}\right] \tag{5.42}$$

with k_R^s being the velocity of surface remodeling when the bone is in the disuse state.

5.6.3 Numerical Assessments

In the numerical analysis, three cases are considered. The axial force is assumed to be $P = 1.0$ kN in all cases.

5.6.3.1 Effect of Pin Size on Bone Remodeling

To study the relationship between pin size and bone porosity, three sizes of the pin δ/a are considered: $\delta/a = 3\%$, 5%, and 6%. The Young's modulus of the pin is $E = 150$ GPa. Figure 5.19 shows the results of bone porosity as a function of t for the three sizes of the pin above.

Figure 5.20 shows the effect of pin size on bone surface remodeling. It can be seen from Figures 5.19 and 5.20 that the pin size has a significant effect on the bone remodeling process. When the pin size is relatively small—say, $\delta/a = 3\%$—bone porosity increases significantly (or bone density decreases) along with an increase in the remodeling time. The explanation for this is as follows. When the pin size is small, the contact force between bone tissue and the pin is also relatively small, which causes the bone to be in a state of near disuse. Then the value of internal growth factors may be smaller than the threshold value, activating the remodeling process for the bone tissues that are in disuse. This process leads to an increase in bone porosity and in the inner radius of the bone. However, when the pin size increases to $\delta/a = 5\%$ or above, the bone porosity seems to be kept in a constant state as time progresses. The change in the inner radius of the bone is slowed down significantly. Perhaps the contact force between the pin and the bone just causes the bone growth factor to be in a state of equilibrium.

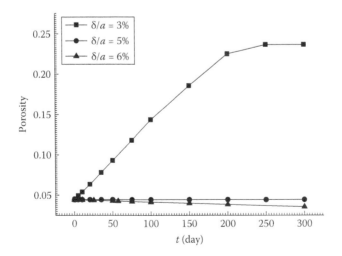

FIGURE 5.19
Bone porosity versus time t for three sizes of pin.

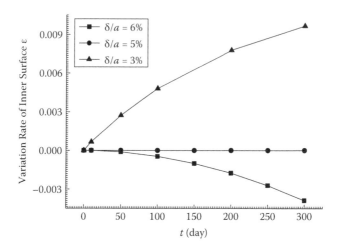

FIGURE 5.20
Effect of pin size on surface bone remodeling.

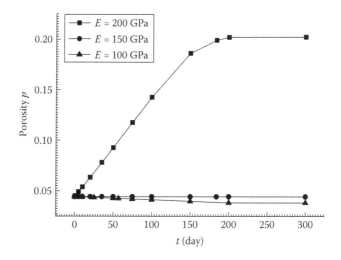

FIGURE 5.21
Effect of pin stiffness on internal bone remodeling.

5.6.3.2 Effect of Pin Stiffness on Bone Remodeling

To study the effect of Young's modulus of the pin on the bone remodeling process, the following values of Young's modulus of the pin are considered: $E = 100, 150,$ and 200 GPa. The size of the pin is defined as $\delta/a = 5\%$. Figure 5.21 shows the results of bone porosity as a function of time t for the three Young's moduli given here. Figure 5.22 shows the corresponding results of variation of the inner radius with the time t. It is evident from Figure 5.21 that, when

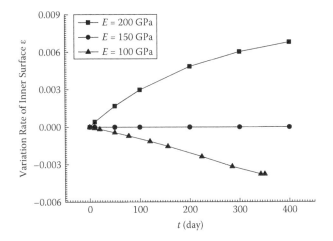

FIGURE 5.22
Effect of pin stiffness on surface bone remodeling.

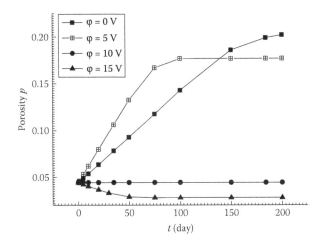

FIGURE 5.23
Effect of electromagnetic field on internal bone remodeling.

$E = 200$ GPa, the porosity increases significantly as time progresses, whereas the bone porosity remains nearly constant over time when $E = 150$ GPa or below. Figure 5.22 also shows that the inner radius increases quickly with time t when $E = 200$ GPa, but the inner radius a remains constant for $E = 150$ Gpa, and a decreases slowly with time t for $E = 100$ GPa.

5.6.3.3 Effect of Electromagnetic Field on Bone Remodeling

Assuming that $\varphi = 0, 5, 10, 15$ V; $\psi = 1$ A; $f_e = 2$ Hz; $\delta/a = 5\%$; and $E = 100$ Gpa, Figure 5.23 records variation of porosity with time t under the loading

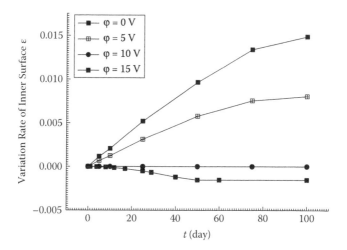

FIGURE 5.24
Effect of electromagnetic field on surface bone remodeling.

described. Figure 5.24 lists the corresponding variation of inner radius with time t. It can be seen from Figures 5.23 and 5.24 that an electromagnetic field also has a significant effect on the bone remodeling process. When the applied electric potential is relatively small—say, $\varphi = 0$ or 5 V—both the bone porosity and inner radius can increase quickly with the time t, whereas the bone porosity remains nearly constant when the applied electric potential increases to $\varphi = 10$ V and above. The variation of inner radius has a similar trend to that of bone porosity.

References

1. Hazelwood S. J., Martin R. B., Rashid M. M., Rodrigo J. J. A mechanistic model for internal bone remodeling exhibits different dynamic responses in disuse and overload. *Journal of Biomechanics* 34 (3): 299–308 (2001).
2. Qu C. Y., Qin Q. H., Kang Y. L. A hypothetical mechanism of bone remodeling and modeling under electromagnetic loads. *Biomaterials* 27 (21): 4050–4057 (2006).
3. He X. Q., Qu C. Y., Qin Q. H. A theoretical model for surface bone remodeling under electromagnetic loads. *Archive of Applied Mechanics* 78 (3): 163–175 (2008).
4. Frost H. M. Changing concepts in skeletal physiology: Wolff's law, the Mechanostat, and the "Utah paradigm." *American Journal of Human Biology* 10 (5): 599–605 (1998).
5. Wolff J. *Das gesetz der transformation der knochen.* Berlin: A. Hirschwald (1892).
6. Cowin S. C. Adaptive elasticity. *Bulletin of the Australian Mathematical Society* 26 (1): 57–80 (1982).

7. Gjelsvik A. Bone remodeling and piezoelectricity—I. *Journal of Biomechanics* 6 (1): 69–77 (1973).
8. Gjelsvik A. Bone remodeling and piezoelectricity—II. *Journal of Biomechanics* 6 (2): 187–193 (1973).
9. Martin R. B., Burr D. B. A hypothetical mechanism for the stimulation of osteonal remodeling by fatigue damage. *Journal of Biomechanics* 15 (3): 137–139 (1982).
10. Beaupre G. S., Orr T. E., Carter D. R. An approach for time-dependent bone modeling and remodeling—Theoretical development. *Journal of Orthopaedic Research* 8 (5): 651–661 (1990).
11. Beaupre G. S., Orr T. E., Carter D. R. An approach for time-dependent bone modeling and remodeling—Application—A preliminary remodeling simulation. *Journal of Orthopaedic Research* 8 (5): 662–670 (1990).
12. Qin Q. H., Ye J. Q. Thermoelectroelastic solutions for internal bone remodeling under axial and transverse loads. *International Journal of Solids and Structures* 41 (9–10): 2447–2460 (2004).
13. Carter D. R., Orr T. E., Fyhrie D. P. Relationships between loading history and femoral cancellous bone architecture. *Journal of Biomechanics* 22 (3): 231–244 (1989).
14. Hart R. T., Davy D. T., Heiple K. G. A computational method for stress analysis of adaptive elastic materials with a view toward applications in strain-induced bone remodeling. *Journal of Biomechanical Engineering—Transactions of the ASME* 106 (4): 342–350 (1984).
15. Huiskes R., Weinans H., Grootenboer H. J., Dalstra M., Fudala B., Slooff T. J. Adaptive bone-remodeling theory applied to prosthetic-design analysis. *Journal of Biomechanics* 20 (11–12): 1135–1150 (1987).
16. Prendergast P. J., Taylor D. Prediction of bone adaptation using damage accumulation. *Journal of Biomechanics* 27 (8): 1067–1076 (1994).
17. Turner C. H., Anne V., Pidaparti R. M. V. A uniform strain criterion for trabecular bone adaptation: Do continuum-level strain gradients drive adaptation? *Journal of Biomechanics* 30 (6): 555–563 (1997).
18. Bassett C. A. L., Valdes M. G., Hernandez E. Modification of fracture repair with selected pulsing electromagnetic fields. *Journal of Bone and Joint Surgery, American* vol. 64 (6): 888–895 (1982).
19. McLeod K. J., Rubin C. T. The effect of low-frequency electrical fields on osteogenesis. *Journal of Bone and Joint Surgery, American* vol. 74A (6): 920–929 (1992).
20. Giordano N., Battisti E., Geraci S., Fortunato M., Santacroce C., Rigato M., Gennari L., Gennari C. Effect of electromagnetic fields on bone mineral density and biochemical markers of bone turnover in osteoporosis: A single-blind, randomized pilot study. *Current Therapeutic Research—Clinical and Experimental* 62 (3): 187–193 (2001).
21. Korenstein R., Somjen D., Fischler H., Binderman I. Capacitative pulsed electric-stimulation of bone-cells—Induction of cyclic amp changes and DNA synthesis. *Biochimica et Biophysica Acta* 803 (4): 302–307 (1984).
22. Luben R. A., Cain C. D., Chen M. C. Y., Rosen D. M., Adey W. R. Effects of electromagnetic stimuli on bone and bone cells in vitro—inhibition of responses to parathyroid hormone by low-energy low-frequency fields. *Proceedings of National Academy of Sciences of the United States of America—Biological Sciences* 79 (13): 4180–4184 (1982).

23. Friedenb, Zb., Zemsky L. M., Pollis R. P., Brighton C. T. Response of nontraumatized bone to direct current. *Journal of Bone and Joint Surgery, American* vol. A 56 (5): 1023–1030 (1974).
24. Lavine L. S., Lustrin I., Shamos M. H., Moss M. L. Influence of electric current on bone regeneration in vivo. *Acta Orthopaedica Scandinavica* 42 (4): 305–314 (1971).
25. Brighton C. T., Friedenberg Z. B., Zemsky L. M., Pollis P. R. Direct-current stimulation of non-union and congenital pseudarthrosis—Exploration of its clinical application. *Journal of Bone and Joint Surgery, American* vol. A 57 (3): 368–377 (1975).
26. Sharrard W. J. W., Sutcliffe M. L., Robson M. J., Maceachern A. G. The treatment of fibrous non-union of fractures by pulsing electromagnetic stimulation. *Journal of Bone and Joint Surgery, British* vol. 64 (2): 189–193 (1982).
27. Lammens J., Liu Z. D., Aerssens J., Dequeker J., Fabry G. Distraction bone healing versus osteotomy healing: A comparative biochemical analysis. *Journal of Bone Mineral Resource* 13 (2): 279–286 (1998).
28. Weinreb M., Suponitzky I., Keila S. Systemic administration of an anabolic dose of PGE(2) in young rats increases the osteogenic capacity of bone marrow. *Bone* 20 (6): 521–526 (1997).
29. Fitzsimmons R. J., Strong D. D., Mohan S., Baylink D. J. Low-amplitude, low-frequency electric field-stimulated bone cell proliferation may in part be mediated by increased igf-ii release. *Journal of Cellular Physiology* 150 (1): 84–89 (1992).
30. Nagai M., Ota M. Pulsating electromagnetic field stimulates messenger RNA expression of bone morphogenetic protein-2 and protein-4. *Journal of Dental Research* 73 (10): 1601–1605 (1994).
31. Zhuang H. M., Wang W., Seldes R. M., Tahernia A. D., Fan H. J., Brighton C. T. Electrical stimulation induces the level of TGF-beta 1 mRNA in osteoblastic cells by a mechanism involving calcium/calmodulin pathway. *Biochemical and Biophysical Research Communications* 237 (2): 225–229 (1997).
32. Fukada E., Yasuda I. On the piezoelectric effect of bone. *Journal of the Physical Society of Japan* 12 (10): 1158–1162 (1957).
33. Qin Q. H. Multifield bone remodeling under axial and transverse loads. In *New research on biomaterials,* ed. Boomington D. R., pp. 49–91. New York: Nova Science Publishers (2007).
34. Kelly P. J., Bronk J. T. Venous pressure and bone formation. *Microvascular Research* 39 (3): 364–375 (1990).
35. Lanyon L. E. Functional strain as a determinant for bone remodeling. *Calcified Tissue International* 36:S56–S61 (1984).
36. Turner C. H. Three rules for bone adaptation to mechanical stimuli. *Bone* 23 (5): 399–407 (1998).
37. Frost H. M. Bone's Mechanostat: A 2003 update. *Anatomical Record Part A—Discoveries in Molecular Cellular and Evolutionary Biology* 275A (2): 1081–1101 (2003).
38. Frost H. M. Bone mass and the Mechanostat—A proposal. *Anatomical Record* 219 (1): 1–9 (1987).
39. Frost H. M. Perspectives: A proposed general model of the "mechanostat" (suggestions from a new skeletal-biologic paradigm). *Anatomical Record* 244 (2): 139–147 (1996).
40. Rubin C. T., Lanyon L. E. Regulation of bone mass by mechanical strain magnitude. *Calcified Tissue International* 37 (4): 411–417 (1985).

41. Jee W. S. S., Li X. J., Ke H. Z. The skeletal adaptation to mechanical usage in the rat. *Cells and Materials,* Supplement 1, 131–142 (1991).
42. Batra G. S., Hainey L., Freemont A. J., Andrew G., Saunders P. T. K., Hoyland J. A., Braidman I. P. Evidence for cell-specific changes with age in expression of estrogen receptor (ER) alpha and beta in bone fractures from men and women. *Journal of Pathology* 200 (1): 65–73 (2003).
43. Skerry T. M., Bitensky L., Chayen J., Lanyon L. E. Early strain-related changes in enzyme-activity in osteocytes following bone loading in vivo. *Journal of Bone Mineral Research* 4 (5): 783–788 (1989).
44. Kleinnulend J., Vanderplas A., Semeins C. M., Ajubi N. E., Frangos J. A., Nijweide P. J., Burger E. H. Sensitivity of osteocytes to biomechanical stress in vitro. *FASEB Journal* 9 (5): 441–445 (1995).
45. Mullender M. G., Huiskes R. Proposal for the regulatory mechanism of Wolff's law. *Journal of Orthopaedic Research* 13 (4): 503–512 (1995).
46. Nielsen H. M., Andreassen T. T., Ledet T., Oxlund H. Local injection of TGF-beta increases the strength of tibial fractures in the rat. *Acta Orthopaedica Scandinavica* 65 (1): 37–41 (1994).
47. Martin R. B. A theory of fatigue damage accumulation and repair in cortical bone. *Journal of Orthopaedic Research* 10 (6): 818–825 (1992).
48. Currey J. D. The effect of porosity and mineral-content on the Young's modulus of elasticity of compact bone. *Journal of Biomechanics* 21 (2): 131–139 (1988).
49. Rho J. Y., Ashman R. B., Turner C. H. Young's modulus of trabecular and cortical bone material—Ultrasonic and microtensile measurements. *Journal of Biomechanics* 26 (2): 111–119 (1993).
50. Martin R. B: The usefulness of mathematical-models for bone remodeling. *Yearbook of Physical Anthropology* 28, 227–236 (1985).
51. Albright J., Skinner H. Bone: Structural organization and remodeling dynamics. In *The scientific basis of orthopaedics,* 2nd ed., ed. Albright J., Brand R., pp. 161–198 East Norwalk, CT: Appleton and Lange (1987).
52. Brockstedt H., Bollerslev J., Melsen F., Mosekilde L. Cortical bone remodeling in autosomal dominant osteopetrosis: A study of two different phenotypes. *Bone* 18 (1): 67–72 (1996).
53. Parfitt A. M. Osteonal and hemi-osteonal remodeling: The spatial and temporal framework for signal traffic in adult human bone. *Journal of Cellular Biochemistry* 55 (3): 273–286 (1994).
54. Whalen R. T., Carter D. R., Steele C. R. Influence of physical activity on the regulation of bone density. *Journal of Biomechanics* 21 (10): 825–837 (1988).
55. Bentolila V., Boyce T. M., Fyhrie D. P., Drumb R., Skerry T. M., Schaffler M. B. Intracortical remodeling in adult rat long bones after fatigue loading. *Bone* 23 (3): 275–281 (1998).
56. Mori S., Burr D. B. Increased intracortical remodeling following fatigue damage. *Bone* 14 (2): 103–109 (1993).
57. Burr D. B., Martin R. B. Calculating the probability that microcracks initiate resorption spaces. *Journal of Biomechanics* 26 (4–5): 613–616 (1993).
58. Schaffler M. B., Choi K., Milgrom C. Aging and matrix microdamage accumulation in human compact bone. *Bone* 17 (6): 521–525 (1995).
59. Brockstedt H., Kassem M., Eriksen E. F., Mosekilde L., Melsen F. Age-related and sex-related changes in iliac cortical bone mass and remodeling. *Bone* 14 (4): 681–691 (1993).

60. Lanyon L. E., Hampson W. G. J., Goodship A. E., Shah J. S. Bone deformation recorded in vivo from strain gauges attached to human tibial shaft. *Acta Orthopaedica Scandinavica* 46 (2): 256–268 (1975).
61. Rubin C. T., Lanyon L. E. Regulation of bone formation by applied dynamic loads. *Journal of Bone and Joint Surgery,* American vol. 66A (3): 397–402 (1984).
62. Carter D. R., Fyhrie D. P., Whalen R. T. Trabecular bone-density and loading history—Regulation of connective-tissue biology by mechanical energy. *Journal of Biomechanics* 20 (8): 785–794 (1987).
63. Ruimerman R. *Modeling and remodeling in bone tissue.* Eindhoven, the Netherlands: University Press Facilities (2005).
64. Qu C. Y. Simulation of bone remodeling under multifield loadings. PhD thesis, Tianjin University (2007).
65. Hillsley M. V., Frangos J. A. Bone tissue engineering—The role of interstitial fluid flow—Review. *Biotechnology and Bioengineering* 43 (7): 573–581 (1994).
66. Qu C. Y., Qin Q. H., Kang Y. L. Theoretical prediction of surface bone remodeling under axial and transverse loads. 9th International Conference on Inspection Appraisal Repairs & Maintenance of Structures (2005).

6

Effect of Parathyroid Hormone on Bone Metabolism

6.1 Introduction

In Chapters 3 and 4 we discussed internal and surface bone remodeling affected by temperature change and mechanical and electrical loading based on the concept of bone density. Bone remodeling processes induced by PTH, mechanical loading, and PEMF at the cellular level—based on the concept of the RANK–RANKL–OPG pathway—are described in this and the subsequent two chapters. The RANK–RANKL–OPG pathway involves three major components [1]:

1. The receptor activator of nuclear factor kappa B (NF-κB) (RANK) expressed on the surface of hemopoietic precursor cells (also referred to as osteoclast precursor cells)
2. RANKL, a polypeptide found on the surface of osteoblastic cells and proteolytically released in soluble form
3. OPG, a "decoy receptor" molecule released by osteoblastic cells

Differentiation and activation of osteoclast precursor cells into mature (active) osteoclasts requires binding of RANKL to RANK. The RANK–RANKL interaction is inhibited by OPG, which binds to RANKL. We begin in this chapter with the introduction of a mathematical model for simulating the anabolic behavior of bone affected by PTH. The model incorporates a new understanding of the interaction of PTH and other factors with the RANK–RANKL–OPG pathway into bone remodeling, which is able to simulate anabolic actions of bone induced by PTH at the cellular level. The mathematical model described here provides a detailed biological description of bone remodeling using the latest experimental findings and can explain the mechanism of the bone anabolic action by PTH that is administered intermittently as well as the catabolic effect when it is applied continuously.

Using the mathematical model, an investigation is conducted into the effects of intermittent PTH application on the site of bone remodeling at three different stages: the beginning, middle, and end of remodeling. The results show that, because of the negative effect on bone mass at the beginning of PTH application, which occurs in all PTH therapies, it is optimum therapy that PTH be administered at the beginning of the remodeling process in order to decrease the negative effect and obtain more bone gain. The development of this conclusion provides a reasonable basis for supporting the design of optimal dosing strategies for PTH-based anti-osteoporosis treatments.

As indicated in references 2 and 3 and Chapter 1, bone is a highly specialized support tissue that is characterized by its rigidity and hardness. It is composed of support cells (osteoblasts and osteocytes), remodeling cells (osteoclasts), a nonmineral matrix of collagen and noncollagenous proteins (osteoid), and inorganic mineral salts deposited within the matrix. The major functions of bone are (a) to provide structural support for the body and protection of vital organs, (b) to provide an environment for marrow (both blood forming and fat storage), and (c) to act as a mineral reservoir for calcium homeostasis in the body.

It is well known that a bone is a living organ that undergoes remodeling throughout life. Remodeling involves a coordinated action of a number of cells that work in concert in BMUs. In the remodeling process, bone is destroyed or resorbed by osteoclasts and then laid down by osteoblasts. The life span of a single BMU is about 6–9 months, during which several generations of osteoclasts (average life of about 2 weeks) and osteoblasts (average life of about 3 months) are formed. Remodeling results from the action of osteoblasts and osteoclasts, which are the two principal cell types found in bones, and defects such as microfractures are repaired by their coupling reaction. Osteoblasts produce the matrix that becomes mineralized in a well-regulated manner. This mineralized matrix can be removed by the activity of osteoclasts when activated. In a homeostatic equilibrium, resorption and formation are balanced so that old bone tissues are continuously replaced by new tissue regulated by a variety of biochemical and mechanical factors. In 1963 Frost defined this phenomenon as bone remodeling [4].

In normal adults there is a balance between the amount of bone resorbed by osteoclasts and the amount formed by osteoblasts [5]. In this complex process, a bone is remodeled by groups of cells derived from different BMUs [6]. The remodeling cycle consists of three consecutive phases [7]: resorption, reversal, and formation. Resorption begins with the migration of partially differentiated mononuclear preosteoclasts to the bone surface, where they form multinucleated osteoclasts. After the completion of osteoclastic resorption, there is a reversal phase when mononuclear cells appear on the bone surface. These cells prepare the surface for new osteoblasts to begin bone formation and provide signals for osteoblast differentiation and migration. The formation phase follows, with osteoblasts laying down bone until the resorbed bone is completely replaced by new bone. When this phase is completed, the surface is covered with flattened lining cells and a prolonged resting period begins until a new remodeling cycle is initiated. The stages of the remodeling cycle have different lengths. Resorption probably continues for about 2 weeks; the reversal phase may last up to 4 or 5 weeks, and formation can continue for 4 months until the new bone structural unit is completely created.

It is noted that a bone, the major reservoir of body calcium, is under the hormonal control of PTH [8]. PTH binds to cells of the osteoblast lineage [9] and produces both anabolic and catabolic effects. The fact that PTH has dual effects depending on its administration method raises important questions about the mechanisms of its action on bone formation and resorption. The clinical trial reported in Neer et al. [10] clearly demonstrated that intermittent exposure of bone to PTH can increase bone formation and bone mass in humans. Jilka et al. [11] further reported that daily PTH injections in mice with either normal bone mass or osteopenia due to defective osteoblastogenesis increased bone formation without affecting the generation of new osteoblasts. This is in marked contrast to continuous PTH exposure, which causes net bone loss (resorption) [12,13].

In summary, the overall effect of PTH on bone mass depends primarily on its mode of administration. Whereas a continuous increase in PTH levels decreases bone mass, intermittent PTH administration has an anabolic action on a bone [14]. It has been recognized that PTH is the most important regulator of calcium homeostasis in the bone remodeling process. It maintains serum calcium concentrations by stimulating bone resorption and increasing renal tubular calcium reabsorption and renal calcitriol production. PTH is secreted in response to a drop in plasma Ca^{2+} levels. With the goal of maintaining plasma Ca^{2+}, PTH increases bone resorption to release Ca^{2+} stored in bones. Acting on osteoblasts, PTH alters the expression of RANKL and OPG (osteoprotegerin), leading to a large increase in the RANKL/OPG ratio, thus stimulating osteoclastogenesis and bone resorption [15,16]. RANKL here is the abbreviation for receptor activator for nuclear factor kappa-B ligand, found on the surface of stromal cells, osteoblasts, and T-cells. RANKL is a member of the tumor necrosis factor (TNF) cytokine family, which is a ligand for osteoprotegerin and functions as a key factor for osteoclast differentiation and activation.

Mathematical modeling provides a powerful tool to predict the net outcome of multiple, simultaneous actions of autocrine, paracrine, and endocrine factors on the process of bone remodeling. Although only few attempts have been made to reconstruct the process of bone remodeling mathematically at a cellular level, there is increasing interest in this approach:

- Kroll [8] and Rattanakul et al. [17] each proposed a mathematical model accounting for the differential activity of PTH administration on bone accumulation.

- Komarova et al. [18] presented a theoretical model of autocrine and paracrine interactions among osteoblasts and osteoclasts.
- Komarova [19] also developed a mathematical model describing the actions of PTH at a single site of bone remodeling, where osteoblasts and osteoclasts are regulated by local autocrine and paracrine factors.
- Potter et al. [20] proposed a mathematical model for PTH receptor (PTH1R) kinetics, focusing on the receptor's response to PTH dosing to discern bone formation responses from bone resorption.
- Lemaire et al. [21] incorporated detailed biological information and a RANK–RANKL–OPG pathway into the remodeling cycle; however, only the catabolic effect of PTH on a bone is included in that model.
- Pivonka et al. [22] developed an extended bone-cell population model based on the work of Lemaire et al. [21] to explore the model structure of cell–cell interactions theoretically and then investigated the role of the RANK–RANKL–OPG system in bone remodeling [1].
- Incorporating the latest experimental findings and mathematical advances, Wang, Qin, and Kalyanasundaram [2] developed a mathematical model of bone cell population dynamics that could simulate the anabolic behavior of bone affected by intermittent administration of PTH, and that model was further detailed in [2].

It is expected that the model concerning systemic and local regulation of bone remodeling will lead to new approaches in the diagnosis and treatment of skeletal disorders. In particular, this model will help to develop new therapeutic approaches at the molecular and cellular level based on the definition of abnormalities of the osteoblastic and osteoclastic lineage that lead to bone diseases such as osteoporosis. Emphasis in this chapter, however, is given to the developments in references 2 and 21.

6.2 Structure of the Model and Assumption

The overall integrity of bone appears to be controlled by hormones and many other proteins secreted by both hemopoietic bone marrow cells and bone cells. There is both systemic and local regulation of bone cell function. As stated in Chapter 1, PTH is the most important regulator of calcium homeostasis, which can stimulate bone formation when given intermittently and bone resorption when delivered continuously [23]. Moreover, PTH is currently involved in numerous clinical trials as an anabolic agent for the treatment of low bone mass in osteoporosis. Forteo (PTH 1-34) has been approved as an anabolic therapy by the United States FDA [10,24]. The IGF system is also important for skeleton growth. It is among the major determinants of adult bone mass through its effect on the regulation of both bone formation and resorption [25]. IGF-1 promotes chondrocyte and osteoblast differentiation and growth. It is also a pivotal factor in the coupling of bone turnover because it is stored in the skeletal matrix and released during bone resorption [26] and stimulates bone formation directly.

PTH receptors are largely expressed on the osteoblastic surface [27,28]. By binding these receptors, quasisteady-state levels of plasma PTH stimulate the production of RANKL and inhibit the production of OPG by osteoblasts [28–30], which causes an increase in activated osteoclast (AOC) numbers. A direct effect of PTH on osteoblasts that are anti-apoptosis has also been experimentally observed [27,31].

As far as the local regulation of bone cell function is concerned, after the recent discovery of the RANK–RANKL–OPG system, there is a clearer picture regarding the control of osteoclastogenesis and bone remodeling in general. The main switch for osteoclastic bone resorption is the RANKL [32], a cytokine that is released by activated osteoblasts. Its action on the RANK receptor is regulated by OPG, a decoy receptor, which is also derived from osteoblastic lineage preosteoblasts. Osteoclast-to-osteoblast cross talk occurs mostly through growth factors, such as TGF-β, which are released from the bone matrix during resorption.

The opposite phenotypes of OPG overexpression or with RANKL deletion mice (osteopetrosis) and OPG-deficient or with RANKL overexpression (osteoporosis) have led to the hypothesis that OPG and RANKL can be the mediators for the stimulatory or inhibitory effects of a variety of systemic hormones, growth factors, and cytokines on osteoclastogenesis [7]. This has recently been referred to as "the convergence hypothesis" in that the activity of the resorptive and antiresorptive agents "converges" at the level of these two mediators, whose final ratio controls the degree of osteoclast differentiation, activation, and apoptosis [33].

The previously described BMU comprises a collection of different cell types with different origins. The osteoclast teams that line the cutting cone are derived from hematopoietic stem cells residing mainly in the marrow and spleen. Osteoclastogenesis begins when a hematopoietic stem cell is stimulated to generate mononuclear cells, which then become committed preosteoclasts and are introduced into the blood stream. This step requires expression of the *ETS* (*E-twenty six*, one of the largest families of transcription factors and unique to metazoans) family transcription factor PU.1 and macrophage colony stimulating factor (M-CSF) [34,35]. The circulating precursors exit the peripheral circulation at or near the site to be resorbed, fusing with one another to form a multinucleated immature osteoclast. Fusion of the mononuclear cells into an immature osteoclast requires the presence of M-CSF and RANKL, a tumor necrosis factor family member [36]. RANKL interacts with a receptor on an osteoclast precursor called RANK. Further differentiation of the immature osteoclast into mature and active osteoclasts occurs only under the continued presence of RANKL [37]. The RANK–RANKL interaction results in activation, differentiation, and fusion of hematopoietic cells of the osteoclast lineage so that they begin the process of resorption. Further, it prolongs osteoclast survival by suppressing apoptosis [38].

Osteoblast development follows a different course, beginning with the local proliferation of mesenchymal stem cells residing in the marrow, which can also give rise to other types of cells such as myocytes, chondrocytes, and adipocytes [39]. Proliferating precursors are pushed toward the preosteoblasts—responding osteoblasts (ROBs)—under the complex effects of specific factors such as PTH and TGF-β [39]. After further differentiation, ROBs mature to active osteoblasts (AOBs), which are responsible for bone formation. Eventually, osteoblasts die or transform to either lining cells or osteocytes [40].

The logical structure of the model is presented in Figure 6.1, which shows the simplified lineages of osteoblasts, osteoclasts, and their interactions. See references 2 and 3 for more details of this model.

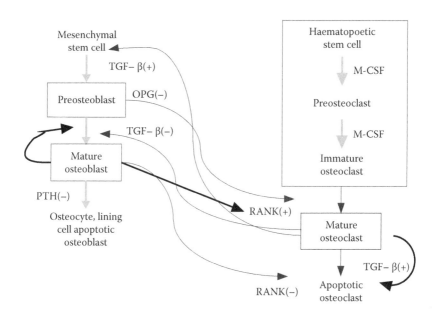

FIGURE 6.1
Schematic representation of structure of the model. The solid arrow with a (+) or (−) beside it represents a stimulatory or inhibitory action by the factor. The thin frame squares indicate types of cells that are included in this model.

6.3 Bone Remodeling Formulation

In this section, formulations presented in references 2 and 21 are briefly reviewed. In the model shown in Figure 6.1, cellular interactions are carried out via activation of cell receptors. The receptors either bind molecules secreted by other cell types called paracrine, or with molecules secreted by the same cell called autocrine, or with other transmembrane molecules via direct cell-to-cell contacts. The different cell types represented in the model respond to the activation of their receptors by producing new molecules, differentiating, or dying [21]. The mathematical formulation of the model is primarily influenced by physiological events involving receptor binding and intracellular signaling modeling [41,42].

Without considering the osteoblastic interactions, the reaction scheme of the binding of PTH with its receptor is represented as follows:

$$\underset{d_p}{\overset{p_p}{\downarrow}} P + P_r \underset{\overrightarrow{k_6}}{\overset{\overleftarrow{k_5}}{}} P_r \bullet P \tag{6.1}$$

where p_p and d_p are abbreviations for PTH production and destruction fluxes.

Applying the law of mass action [21] to reaction equation (6.1), the following ordinary differential equations used to describe the reactions of receptors and corresponding ligands, including PTH (P) with its receptor (P_r), are obtained:

$$\frac{dP}{dt} = S_p + I_p + \left(k_6 \cdot P_r \bullet P - k_5 \left(R_T^P - P_r \bullet P \right) \cdot P \right) \cdot (B + R) - k_P \cdot P \tag{6.2}$$

$$\frac{d(P_r \bullet P)}{dt} = k_5 \left(R_T^P - P_r \bullet P \right) \cdot P - k_6 \cdot P_r \bullet P \tag{6.3}$$

where the small dot stands for the multiplication and the large dot above represents a receptor–ligand complex. For example, $P_r \bullet P$ is the complex formed by PTH and its receptor. The meaning and values of the parameters S_p, I_p, k_5, k_6, and k_p can be found in Table 6.1.

It should be mentioned that, depending on the profile of administration, I_p may be constant or may be time dependent. R_T^P is the number of PTH receptors per cell. This number is supposed to be constant, implying that no significant synthesis, degradation, internalization, or recycling of receptors occurs over the time span for which the model applies. B and R are the concentrations of AOBs and ROBs, respectively. The PTH binding reaction (Equation 6.1) equilibrates much more rapidly than the time it takes for the cell populations of the model to change noticeably. That is also the case for all other binding reactions of the model. Consequently, only the steady states of these molecular events really alter the cell dynamics in the model.

TABLE 6.1

Model Parameters

Symbol	Unit	Valve	Description
C^S	pM	5×10^{-3}	Value of C to obtain half differentiation flux
D_A	Day^{-1}	0.7	Rate of osteoclast apoptosis caused by TGF-β
d_B	Day^{-1}	0.7	Differentiation rate of responding osteoblasts
D_C	pM Day^{-1}	2.1×10^{-3}	Differentiation rate of osteoclast precursor
D_R	pM Day^{-1}	7×10^{-4}	Differentiation rate of osteoblast progenitors
f_0	—	0.05	Fixed proportion
I_L	pM Day^{-1}	$0-10^6$	Rate of administration of RANKL
I_o	pM Day^{-1}	$0-10^6$	Rate of administration of OPG
I_P	pM Day^{-1}	$0-10^6$	Rate of administration of PTH
K	pM	10	Fixed concentration of RANK
k_1	pM^{-1} Day^{-1}	10^{-2}	Rate of OPG–RANKL binding
k_2	Day^{-1}	10	Rate of OPG–RANKL unbinding
k_3	pM^{-1} Day^{-1}	5.8×10^{-4}	Rate of RANK–RANKL binding
k_4	Day^{-1}	1.7×10^{-2}	Rate of RANK–RANKL unbinding
k_5	pM^{-1} Day^{-1}	0.02	Rate of PTH binding with its receptor
k_6	Day^{-1}	3	Rate of PTH unbinding
k_B	Day^{-1}	0.189	Rate of elimination of AOB
K_L^P	pmol/pmol cells	3×10^6	Maximum number of RANKL attached on each cell surface
k_O	Day^{-1}	0.35	Rate of elimination of OPG
K_O^P	pmol day^{-1}/ pmol cells	2×10^5	Minimal rate of production of OPG per cell
k_P	Day^{-1}	86	Rate of elimination of PTH
r_L	pM Day^{-1}	10^3	Rate of RANKL production and elimination
S_P	pM Day^{-1}	250	Rate of synthesis of systemic PTH
R^S	pM	5×10^{-2}	Value of R to get half differentiation flux
m_1	% cell^{-1} d^{-1}	122	Average rate of bone resorbed per day per AOC
m_2	% cell^{-1} d^{-1}	195	Average rate of bone formed per day per AOB

Sources: Wang Y. N., Qin Q. H., Kalyanasundaram S. *Molecular & Cellular Biomechanics* 6 (2): 101–112 (2009); Lemaire V. et al. *Journal of Theoretical Biology* 229 (3): 293–309 (2004).

The reaction schemes of the bindings of OPG (O) with RANKL (L) and RANKL with RANK (K) are the following:

$$\overset{p_o}{\underset{d_o}{O}} + \overset{p_L}{\underset{d_L}{L}} \overset{k_1}{\underset{k_2}{\rightleftharpoons}} O \bullet L \quad \text{and} \quad \overset{p_L}{\underset{d_L}{L}} + K \overset{k_3}{\underset{k_4}{\rightleftharpoons}} K \bullet L \tag{6.4}$$

where $O \bullet L$ and $K \bullet L$ represent the OPG–RANKL and RANK–RANKL complexes respectively. The differential equations describing the time evolution of the variables $O, L, O \bullet L,$ and $K \bullet L$ (K is not considered as a model variable; the concentration of RANK is kept fixed to reflect the undiminished availability of the osteoclast precursors) are the following [2,3]:

$$\frac{dO}{dt} = p_O - k_1 \cdot O \cdot L + k_2 \cdot O \bullet L - d_O \qquad (6.5)$$

$$\frac{d(O \bullet L)}{dt} = k_1 \cdot O \cdot L - k_2 \cdot O \bullet L \qquad (6.6)$$

$$\frac{dL}{dt} = p_L - k_1 \cdot O \cdot L + k_2 \cdot O \bullet L - k_3 \cdot K \cdot L + k_4 \cdot K \bullet L - d_L \qquad (6.7)$$

$$\frac{d(K \bullet L)}{dt} = k_3 \cdot K \cdot L - k_4 \cdot K \bullet L \qquad (6.8)$$

where the meaning and values of the parameters $k_1 \sim k_4$ can be found in Table 6.1 on the previous page, and $d_o = k_o \cdot O$ is a linear degradation function, with k_O being the rate of elimination of OPG.

In the analysis, Wang et al. [2] and Wang [3] proposed that cell proliferation rate is proportional to the receptor occupancy [25] and they applied this rule to other types of cell response in addition to cell proliferation. Moreover, the antiproliferative cell responses are inversely proportional to the receptor occupancy. Consequently, the production rate of OPG (p_o) is downregulated by PTH and upregulated by TGF-β, and the expression of p_o is

$$p_O = K_O^P \cdot \left(\frac{1}{\pi_P} + \pi_C \right) R + I_O \qquad (6.9)$$

where K_O^P is the minimal production rate of OPG per cell, I_O is a rate of external injection of OPG, and at equilibrium state the proportion of occupied PTH receptor π_P is

$$\pi_P = \sqrt{\frac{P_r \bullet P}{R_T^P}} = \sqrt{\frac{P}{P + P^s}} \qquad (6.10)$$

Considering $P^s = k_6 / k_5, P = (S_P + I_P) / k_P = P^0 + \bar{P}$, with $P^0 = S_P/k_P$ and $\bar{P} = S_P/k_P$, at equilibrium state, we have

$$\pi_P \approx \sqrt{\frac{\bar{P} + P^0}{\bar{P} + P^s}} = \sqrt{\frac{I_P/k_P + S_P/k_P}{I_P/k_P + k_6/k_5}} \qquad (6.11)$$

for $P^0 \ll P^s$.

As mentioned in Lemaire et al. [21], unlike OPG, RANKL is attached to osteoblast surfaces and occupies a restricted space. The surface of osteoblasts can carry only a limited number of RANKL molecules. Therefore, the RANKL production rate p_L should be self-limiting to prevent the number of RANKL from exceeding the carrying capacity of the osteoblast surface. The effective carrying capacity is imposed by the cell response to PTH binding, although there is a possible maximum number of RANKL attached on each cell surface, noted as K_L^P. Applying the same rule as that for obtaining P_O, the control of RANKL concentration on the cell surface is obtained as [21]

$$p_L - d_L = r_L \cdot \left(1 - \frac{L + O \bullet L + K \bullet L}{K_L^P \cdot \pi_P \cdot B} \right) + I_L \tag{6.12}$$

where r_L is the rate of RANKL production and elimination, and I_L is a rate of external injection of RANKL. $L + O \bullet L + K \bullet L$ is the current total concentration of RANKL, where the denominator corresponds to the imposed carrying capacity. In Equation (6.12), the concentration of RANKL that the osteoblasts aim to produce is governed by the value of π_P: $K_L^P \cdot \pi_P \cdot B$. If the total concentration of RANKL is below this value, the rate of production of RANKL is positive; otherwise, it is negative. Pseudosteady states of the RANKL and OPG concentrations are calculated from Equations (6.5)–(6.9) and (6.12):

$$L = \frac{K_L^P \cdot \pi_P \cdot B}{1 + k_3 \, K/k_4 + k_1 \, O/k_2} \cdot \left(1 + \frac{I_L}{r_L} \right) \tag{6.13}$$

$$O = \frac{K_O^P}{k_O \pi_P} R + \frac{I_O}{k_O} \tag{6.14}$$

The osteoblast lineage is supplied from a large population of uncommitted progenitors. These progenitors express a specific TGF-β receptor that, once activated, leads to the differentiation of the progenitors into ROBs. If the rate of release of TGF-β per active osteoclast is constant, the entering flow into the ROB compartment depends only on the mesenchymal stem cells' response to c binding. This response is represented by a proportionality relationship with the TGF-β receptor occupancy π_C:

$$D_R \cdot \pi_C = D_R \cdot \sqrt{\frac{C + C^0}{C + C^s}} \tag{6.15}$$

where
   D_R is a proportionality factor
   C is the concentration of the AOCs
   C^s corresponds to the dissociation binding coefficient of TGF-β with its
      receptor

C^0 ($C^0 = f_0 \cdot C^s$) represents the basal concentration of the AOCs in the system with f_0 being a fixed proportion

The outgoing flow from the ROB compartment is also the feeding flow to the AOB compartment. Under the influence of TGF-β and IGF, which inhibit and stimulate AOB production, respectively,

$$D_B \cdot R \cdot \left(\frac{1}{\pi_C} + \pi_I \right) = D_B \cdot R \cdot \left(\sqrt{\frac{C + C^s}{C + C^0}} + \pi_I \right) \tag{6.16}$$

where D_B is a proportionality factor associated with AOB production.

RANK–RANKL binding promotes the differentiation of mesenchymal stem cells into AOC [2]; the differentiation rate is proportional to the RANK occupancy ratio π_L. This ratio is equal to $K \bullet L/K$ in the case under consideration:

$$\frac{K \bullet L}{K} = \frac{k_3}{k_4} \cdot \frac{K_L^P \pi_P B}{1 + \dfrac{k_3 K}{k_4} + \dfrac{k_1}{k_2 k_O} \cdot \left(\dfrac{K_O^P}{\pi_P} R + I_O \right)} \cdot \left(1 + \frac{I_L}{r_L} \right) \tag{6.17}$$

The flow of differentiation precursors entering the AOC compartment is then given by

$$D_C \cdot \pi_L = D_C \cdot \frac{K \bullet L}{K} \tag{6.18}$$

where D_C is the differentiation rate of the osteoclasts' precursors.

TGF-β induces osteoclast apoptosis via binding to specific receptors and also under the influence of RANKL. This phenomenon is then represented as

$$D_A \cdot (\pi_C - \pi_L) \cdot C = D_A \cdot \frac{C + C^0}{C + C^s} \cdot C \tag{6.19}$$

The equations governing the evolution of the number of cells in each compartment are simply balance equations [21], which means that each cell compartment is fed by an entering flow and is emptied by the outgoing flow of differentiated or apoptotic cells [2]:

$$\frac{dR}{dt} = D_R \cdot \pi_C - D_B \cdot R \cdot \left(\frac{1}{\pi_C} + \pi_I \right) \tag{6.20}$$

$$\frac{dB}{dt} = D_B \cdot R \cdot \left(\frac{1}{\pi_C} + \pi_I \right) - (k_B - \pi_P) \cdot B \tag{6.21}$$

$$\frac{dC}{dt} = D_C \cdot \pi_L - D_A \cdot (\pi_C - \pi_L) \cdot C \tag{6.22}$$

The rate of bone resorption and formation is assumed to be proportional to the numbers of osteoclasts and osteoblasts, respectively:

$$\frac{dZ}{dt} = -m_1 \cdot C + m_2 \cdot B \qquad (6.23)$$

where Z is the total bone mass, and m_1 and m_2 are normalized activities of bone resorption and formation, respectively. Noting that at equilibrium, where the simulation starts, the numbers of AOB, AOC, and ROB do not change with time; solving the following three equations can determine the initial values of B, C, and R:

$$D_R \cdot \pi_C - D_B \cdot R \cdot \left(\frac{1}{\pi_C} + \pi_I \right) = 0 \qquad (6.24)$$

$$D_B \cdot R \cdot \left(\frac{1}{\pi_C} + \pi_I \right) - \left(k_B - \pi_P \right) \cdot B = 0 \qquad (6.25)$$

$$D_C \cdot \pi_L - D_A \cdot \left(\pi_C - \pi_L \right) \cdot C = 0 \qquad (6.26)$$

Values of model parameters and initial conditions of variables are listed in Table 6.1. Model equations (6.20)–(6.23) are then solved using numerical integration by a fourth Runge-Kutta algorithm implemented in Matlab. This is performed in the next section.

6.4 Results and Discussion

To demonstrate the tight coupling between osteoblast and osteoclast, Wang et al. [2] computationally perturbed this system by adding or removing specific cells. The results are displayed from Figures 6.2 to 6.7.

It is evident from Figures 6.2 and 6.3 that adding AOBs can initiate a remodeling cycle from initial stable state as shown in Figure 6.2. Figure 6.2 also shows that the number of AOCs decreases for the first 7 days or so and then increases back to the initial value, whereas the number of AOBs increases as expected; this means that direct administration of AOBs does not have a strong stimulatory effect on AOCs, consistent with experimental observation [2]. Figure 6.3 clearly displays that bone mass increases with administration of AOBs and rises a little more slowly after AOB injection ceases.

Figure 6.4 shows that the administration of AOCs initiates a remodeling cycle and their number remains almost unchanged from approximately the 7th day to the 60th day. There is also a strong and immediate stimulatory

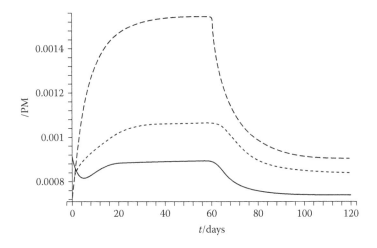

FIGURE 6.2
AOBs are added at a constant rate (0.0001 pM/day) for 60 days from the start. From top to bottom, dashed curves, dotted curves, and solid curves represent AOBs, ROBs, and AOCs, respectively.

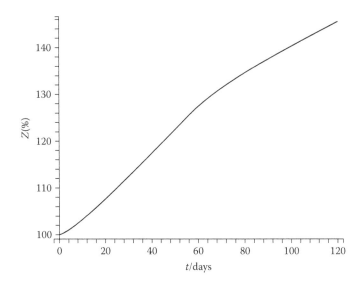

FIGURE 6.3
The effect of adding AOBs at a constant rate (0.0001 pM/day) on bone mass. Changes in bone mass are expressed as a percentage of initial bone mass (100%).

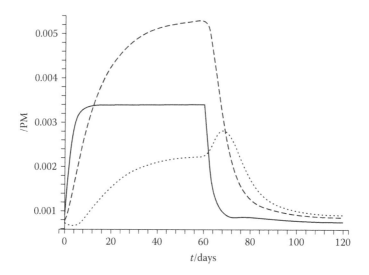

FIGURE 6.4
AOCs are added at a constant rate (0.001 pM/day) for 60 days from the start.

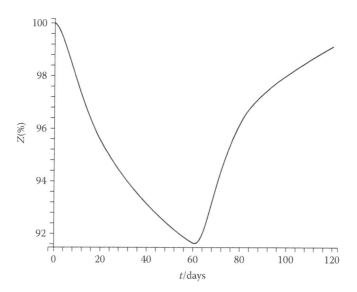

FIGURE 6.5
The effect of adding AOCs at a constant rate (0.001 pM/day) on bone mass.

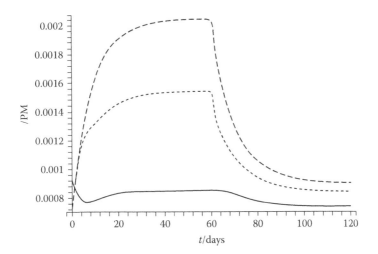

FIGURE 6.6
ROBs are added at a constant rate (0.0001 pM/day) for 60 days from the start.

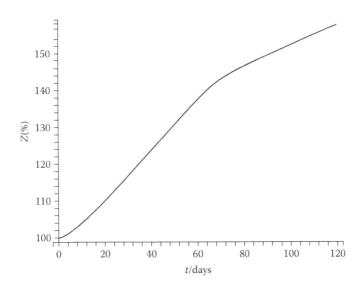

FIGURE 6.7
The effect of adding ROBs at a constant rate (0.0001 pM/day) on bone mass.

effect on ROBs (top dashed curve), which means that the number of ROBs increases immediately with the injection of AOCs and decreases to the initial level after the injection of AOCs ceases. The response of AOBs to the administration of AOCs is to slow down and delay until the injection of AOCs ceases. It can be observed from Figure 6.5 that the amount of AOBs responsible for producing bone mass begins to increase quickly, which accounts for the increase in bone mass after the 60th day.

An interesting phenomenon is observed when ROBs are administered to the system. The AOBs increase in number along with the increase in ROBs, whereas the number of AOCs decreases for approximately the first 7 days and then remains unchanged at a particular level, lower than the initial state, until the 60th day. After 60 days, the number of AOCs equilibrates to an even smaller value, as illustrated in Figure 6.6. Consequently, it is reasonable that the bone mass continues to rise, as shown in Figure 6.7. This observation may have the potential to be exploited therapeutically for metabolic diseases.

The only systemic hormone considered in the model is PTH. As mentioned earlier, intermittent infusion of PTH has potent anabolic effects on bone mass. To test the anabolic action of PTH in the model, the hormone is delivered at a steady rate of 3000 pM/day for 60 days. As can be observed from Figure 6.8, the number of AOCs increases with the infusion of PTH and then drops quickly in response to the cessation of PTH administration. The response of AOBs to the intermittent injection of PTH is relatively slow; the number

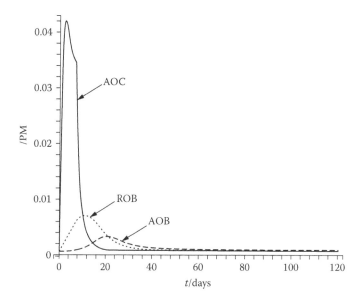

FIGURE 6.8
The responses of AOCs, AOBs, and ROBs to the intermittent administration of PTH for the first 7 days.

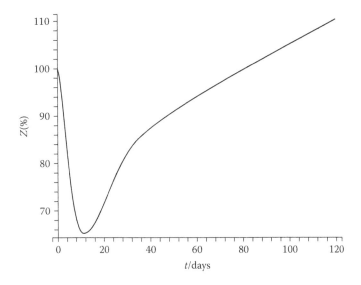

FIGURE 6.9
The effect of intermittent administration of PTH for the first 7 days on bone mass.

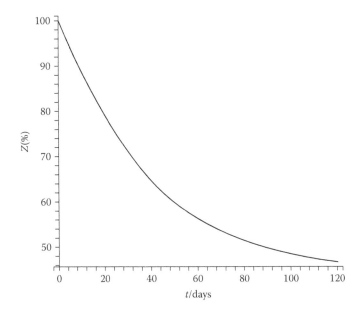

FIGURE 6.10
The effect of continuous administration of PTH for 120 days on bone mass.

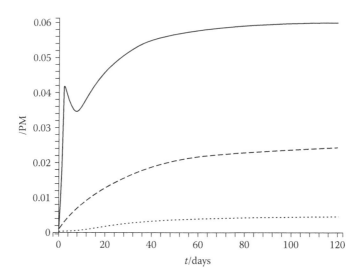

FIGURE 6.11
The responses of AOCs, AOBs, and ROBs to the continuous administration of PTH for 120 days.

of AOBs continues to increase even after the PTH administration ceases, which is the key to the final increase in bone mass (see Figure 6.9). This is in agreement with experimental observation [2].

As expected, under the continuous administration of PTH, bone mass continually decreases, as shown in Figure 6.10. This simulation is in good agreement with experimental results [2]. It can be observed from Figure 6.11 that the responses of AOCs, ROBs, and AOBs are affected by the PTH. In particular, the number of AOCs increases promptly as PTH is injected, followed by a minor drop, and then keeps rising at a slower rate. However, the number of AOBs increases only slightly and at a very slow rate over the first 120 days. Through the direct stimulatory effect of TGF-β released by AOCs, the number of ROBs increases at a higher rate than that of the AOBs.

References

1. Pivonka P., Zimak J., Smith D. W., Gardiner B. S., Dunstan C. R., Sims N. A., Martin T. J., Mundy G. R. Theoretical investigation of the role of the RANK–RANKL–OPG system in bone remodeling. *Journal of Theoretical Biology* 262 (2): 306–316 (2010).
2. Wang Y. N., Qin Q. H., Kalyanasundaram S. A theoretical model for simulating effect of parathyroid hormone on bone metabolism at cellular level. *Molecular & Cellular Biomechanics* 6 (2): 101–112 (2009).

3. Wang Y. N. Mathematical modeling of bone remodeling under mechanical, electromagnetic fields at the cellular level. PhD thesis, Australian National University (2010).

4. Frost H. M. *Bone remodeling dynamics,* vol. 59. Springfield, IL: Charles C Thomas Co. (1963).

5. Robling A. G., Castillo A. B., Turner C. H. Biomechanical and molecular regulation of bone remodeling. *Annual Review of Biomedical Engineering* 8: 455–498 (2006).

6. Parfitt A. M. Osteonal and hemi-osteonal remodeling: The spatial and temporal framework for signal traffic in adult human bone. *Journal of Cellular Biochemistry* 55 (3): 273–286 (1994).

7. Hadjidakis D. J., Androulakis I. I. Bone remodeling. *Annals of the New York Academy of Sciences* 1092 (1): 385–396 (2006).

8. Kroll M. Parathyroid hormone temporal effects on bone formation and resorption. *Bulletin of Mathematical Biology* 62 (1): 163–188 (2000).

9. Datta N. S., Abou-Samra A. B. PTH and PTHrP signaling in osteoblasts. *Cellular Signaling* 21 (8): 1245–1254 (2009).

10. Neer R. M., Arnaud C. D., Zanchetta J. R., Prince R., Gaich G. A., Reginster J. Y., Hodsman A. B., et al. Effect of parathyroid hormone (1-34) on fractures and bone mineral density in postmenopausal women with osteoporosis. *New England Journal of Medicine* 344 (19): 1434–1441 (2001).

11. Jilka R. L., Weinstein R. S., Bellido T., Roberson P., Parfitt A. M., Manolagas S. C. Increased bone formation by prevention of osteoblast apoptosis with parathyroid hormone. *Journal of Clinical Investigation* 104 (4): 439–446 (1999).

12. Tam C. S., Heersche J. N. M., Murray T. M., Parsons J. A. Parathyroid hormone stimulates the bone apposition rate independently of its resorptive action—Differential effects of intermittent and continuous administration. *Endocrinology* 110 (2): 506–512 (1982).

13. Khosla S., Westendorf J. J., Oursler M. J. Building bone to reverse osteoporosis and repair fractures. *Journal of Clinical Investigation* 118 (2): 421–428 (2008).

14. Rubin M. R., Bilezikian J. P. New anabolic therapies in osteoporosis. *Endocrinology & Metabolism Clinics of North America* 32 (1): 285–307 (2003).

15. Lee S. K., Lorenzo J. A. Parathyroid hormone stimulates TRANCE and inhibits osteoprotegerin messenger ribonucleic acid expression in murine bone marrow cultures: Correlation with osteoclast-like cell formation. *Endocrinology* 140 (8): 3552–3561 (1999).

16. Fu Q., Jilka R. L., Manolagas S. C., O'Brien C. A. Parathyroid hormone stimulates receptor activator of NFkappa B ligand and inhibits osteoprotegerin expression via protein kinase A activation of cAMP-response element-binding protein. *Journal of Biological Chemistry* 277 (50): 48868–48875 (2002).

17. Rattanakul C., Lenbury Y., Krishnamara N., Wolwnd D. J. Modeling of bone formation and resorption mediated by parathyroid hormone: Response to estrogen/PTH therapy. *Biosystems* 70 (1): 55–72 (2003).

18. Komarova S. V., Smith R. J., Dixon S. J., Sims S. M., Wahl L. M. Mathematical model predicts a critical role for osteoclast autocrine regulation in the control of bone remodeling. *Bone* 33 (2): 206–215 (2003).

19. Komarova S. V. Mathematical model of paracrine interactions between osteoclasts and osteoblasts predicts anabolic action of parathyroid hormone on bone. *Endocrinology* 146 (8): 3589–3595 (2005).

20. Potter L. K., Greller L. D., Cho C. R., Nuttall M. E., Stroup G. B., Suva L. J., Tobin F. L. Response to continuous and pulsatile PTH dosing: A mathematical model for parathyroid hormone receptor kinetics. *Bone* 37 (2): 159–169 (2005).

21. Lemaire V., Tobin F. L., Greller L. D., Cho C. R., Suva L. J. Modeling the interactions between osteoblast and osteoclast activities in bone remodeling. *Journal of Theoretical Biology* 229 (3): 293–309 (2004).

22. Pivonka P., Zimak J., Smith D. W., Gardiner B. S., Dunstan C. R., Sims N. A., Martin T. J., Mundy G. R. Model structure and control of bone remodeling: A theoretical study. *Bone* 43 (2): 249–263 (2008).

23. Kim C. H., Takai E., Zhou H., von Stechow D., Muller R., Dempster D. W., Guo X. E. Trabecular bone response to mechanical and parathyroid hormone stimulation: the role of mechanical microenvironment. *Journal of Bone Mineral Research* 18 (12): 2116–2125 (2003).

24. Whitfield J. F., Morley P., Willick G. E. Bone growth stimulators. New tools for treating bone loss and mending fractures. *Vitamins and Hormones* 65: 1–80 (2002).

25. Wang J., Zhou J., Cheng C. M., Kopchick J. J., Bondy C. A. Evidence supporting dual, IGF-I-independent and IGF-I-dependent, roles for GH in promoting longitudinal bone growth. *Journal of Endocrinology* 180 (2): 247–255 (2004).

26. Donahue L. R., Rosen C. J. IGFs and bone: The osteoporosis connection revisited. *Proceedings of Society of Experimental Biology and Medicine* 219: 1–7 (1998).

27. Goltzman D. Interactions of PTH and PTHrP with the PTH/PTHrP receptor and with downstream signaling pathways: Exceptions that provide the rules. *Journal of Bone Mineral Research* 14 (2): 173–177 (1999).

28. Teitelbaum S. L. Bone resorption by osteoclasts. *Science* 289 (5484): 1504–1508 (2000).

29. Aubin J. E., Bonnelye E. Osteoprotegerin and its ligand: A new paradigm for regulation of osteoclastogenesis and bone resorption. *Osteoporosis International* 11 (11): 905–913 (2000).

30. Halladay D. L., Miles R. R., Thirunavukkarasu K., Chandrasekhar S., Martin T. J., Onyia J. E. Identification of signal transduction pathways and promoter sequences that mediate parathyroid hormone 1-38 inhibition of osteoprotegerin gene expression. *Journal of Cellular Biochemistry* 84 (1): 1–11 (2002).

31. Isogai Y., Akatsu T., Ishizuya T., Yamaguchi A., Hori M., Takahashi N., Suda T. Parathyroid hormone regulates osteoblast differentiation positively or negatively depending on the differentiation stages. *Journal of Bone Mineral Research* 11 (10): 1384–1393 (1996).

32. Zaidi M. Skeletal remodeling in health and disease. *Nature Medicine* 13 (7): 791–801 (2007).

33. Hofbauer L. C., Khosla S., Dunstan C. R., Lacey D. L., Boyle W. J., Riggs B. L. The roles of osteoprotegerin and osteoprotegerin ligand in the paracrine regulation of bone resorption. *Journal of Bone Mineral Research* 15 (1): 2–12 (2000).

34. Tondravi M. M., McKercher S. R., Anderson K., Erdmann J. M., Quiroz M., Maki R., Teitelbaum S. L. Osteopetrosis in mice lacking haematopoietic transcription factor PU.1. *Nature* 386 (6620): 81–84 (1997).

35. Yoshida H., Hayashi S-I., Kunisada T., Ogawa M., Nishikawa S., Okamura H., Sudo T., Shultz L. D., Nishikawa S-I. The murine mutation osteopetrosis is in the coding region of the macrophage colony stimulating factor gene. *Nature* 345 (6274): 442–444 (1990).

36. Franzoso G., Carlson L., Xing L. P., Poljak L., Shores E. W., Brown K. D., Leonardi A., Tran T., Boyce B. F., Siebenlist U. Requirement for NF-kappa B in osteoclast and B-cell development. *Genes & Development* 11 (24): 3482–3496 (1997).
37. Matsuo K., Galson D. L., Zhao C., Peng L., Laplace C., Wang K. Z. Q., Bachler M. A., et al. Nuclear factor of activated T-cells (NFAT) rescues osteoclastogenesis in precursors lacking c-Fos. *Journal of Biological Chemistry* 279 (25): 26475–26480 (2004).
38. Hsu H., Lacey D. L., Dunstan C. R., Solovyev I., Colombero A., Timms E., Tan H-L., et al. Tumor necrosis factor receptor family member RANK mediates osteoclast differentiation and activation induced by osteoprotegerin ligand. *Proceedings of the National Academy of Sciences* 96 (7): 3540–3545 (1999).
39. Aubin J. E. Bone stem cells. *Journal of Cellular Biochemistry Supplement* 30–31: 73–82 (1998).
40. Jilka R. L., Weinstein R. S., Bellido T., Parfitt A. M., Manolagas S. C. Osteoblast programmed cell death (apoptosis): Modulation by growth factors and cytokines. *Journal of Bone Mineral Research* 13 (5): 793–802 (1998).
41. Knauer D. J., Wiley H. S., Cunningham D. D. Relationship between epidermal growth factor receptor occupancy and mitogenic response. Quantitative analysis using a steady state model system. *Journal of Biological Chemistry* 259 (9): 5623–5631 (1984).
42. Lauffenburger D. A., Linderman J. J. *Receptors. Models for binding, trafficking, and signaling.* New York: Oxford University Press (1996).

7

Cortical Bone Remodeling under Mechanical Stimulus

7.1 Introduction

As an extension of the formulation presented in Chapter 6, this chapter describes a model of bone cell population dynamics for cortical bone remodeling under mechanical stimulus. In the mathematical model presented here, the interstitial fluid shear stress retaining the lacuno-canalicular porosity structure caused by mechanical loading is considered [1,2] as the physical mediator of mechanotransduction by osteocytes, along with the bone mechanosensitivity defined as the function of loading frequency, the number of loading cycles during a loading day, the rest time between loading bouts, and the length of loading period. The extended Hill equation is described to study the case where two ligands bind to their respective receptors on the same cell.

Three rate equations describing changes of osteocytes, nitric oxide (NO), and prostaglandin E2 (PGE_2) are included in this model. A standard bone fracture energy (BFE) is introduced for comprehensively assessing the biomechanical significance of mechanical bone remodeling and is shown to be more appropriate than traditional methods such as bone mineral content (BMC) in terms of experimental results. The results from this model are in good agreement with experimental observations extracted from the literature, including the rapid release of NO after loading, the percentage of BMC increase for different loading regimes, and increased percentage of BFE. The chapter presents a brief description of the dynamics model presented in references 1–4.

At the cellular level, bone remodeling is an organized process where osteoclasts remove old bone and osteoblasts replace old bone with newly formed bone. The osteoclasts and osteoblasts work in a coupled manner within a BMU, which is a mediator mechanism bridging individual cellular activity to whole bone morphology [1,5] that follows an activation–resorption–formation sequence [6].

Bone is a metabolically active tissue capable of adapting its structure and mass to the biological and mechanical environment and repairing damaged sections through remodeling. In particular, mechanical loading has a significant influence on bone remodeling. Disused or reduced loading due to long-term bed rest, casting immobilization, or microgravity conditions (such as experienced by astronauts in a space station or shuttle) induces obvious bone loss and mineral changes [7,8], probably because of a lack of convective fluid flow in the canalicular network. Overuse or increased loading, such as experienced with weightlifting exercises, causes damage to bone tissues, which in turn stimulates bone remodeling and eventually achieves bone gain.

The two important roles of bone remodeling are continuously to replace and repair (1) damaged bone tissues and (2) mineral homeostasis by providing access to stores of calcium and phosphate. Osteoclasts start resorbing bone in response to signals that are as yet unknown but may include direct damage to osteocytes via microcracks in the bone matrix.

The adaptive response of bone to mechanical loading is highly site specific. This is clearly evident at the whole-bone level, with only the bone that is actually loaded undergoing adaptation [9]. This concept is supported by much human research investigating skeletal health indices in athletes, especially in players of racquet sports such as tennis, whose bones of the racquet arm or dominant arm display significantly greater bone mineral density and cortical bone content than in the nonplaying arm [10,11]. The site-specific depositing of new bone is functionally important. The site-specific depositing process puts newly formed bone where it is most required and increases bone strength in the resistible direction of loading, while not necessarily increasing the bone mass or density [12].

Experimental observations [13] show that, in comparison with other organ systems, skeleton tissue is hypocellular and composed primarily of extracellular matrix. Atrabecular bone is a porous latticework of struts or plates of long bones, whereas cortical bone is a dense tissue of low porosity found in the diaphyses of long bones. The microstructure of a cortical bone is organized as a hierarchical arrangement of porosities, including a network of cellular spaces (lacunae), interconnections (canaliculi), and larger vascular (osteonal) canals. This spectrum of microarchitecture features implies the transformation of whole-skeleton loading via localized changes in remodeling cycles.

Current understanding of bone remodeling is primarily based on experimental results in vivo and in vitro. A great deal of research has been carried out on the interactions of autocrine, paracrine, and endocrine activities of receptors and ligands in bone remodeling and mechanotransduction pathways; the role of bone cells involved in this process at cellular and even genic levels; and the influence of mechanical loading on bone formation in bone remodeling. Based on these observations, many hypotheses have been proposed as to the role played by different signaling pathways and the communication between bone cells in bone remodeling. Similarly, many hypotheses have also been reported in the literature regarding bone-cell dynamics of both trabecular and cortical bones under mechanical stimulus. However, due to the complexity of the bone regulation system, which involves numerous factors and interactions, systemic understanding is still incomplete.

Mathematical modeling provides a powerful tool for testing and analyzing various hypotheses in complex systems that are very difficult (such as those that consume time or resources) or just impossible to apply in vivo or in vitro. However, relatively few mathematical models have so far been proposed regarding bone remodeling. In addition to the model mentioned in Chapter 6 for describing the effects of PTH on the bone remodeling process, trabecular bone remodeling has been extensively studied in aspects ranging from prediction of the development of trabecular architecture [14] to the effects of mechanical forces on the maintenance and adaptation of form in trabecular bone [15,16]. Based on the trabecular bone remodeling theory developed by Weinans, Huiskes, and Grootenboer [17], Li [18] developed a new trabecular bone remodeling model that could simulate both the underload and overload resorptions that often occur in dental implant treatments.

However, compared with trabecular bone that represents 20% of the skeletal mass [19], even less theoretical work has been done on cortical bone, which comprises 80% of the skeleton and has high resistance to bending and torsion [19]. The following five representative papers that mathematically analyzed cortical bone remodeling were published in the last decade. The first two [20,21] presented a mathematical model at the cellular level, introducing magnitude of force and number of osteocytes to consider production of NO and PGE_2. The third paper [22] proposed a macroscopic model to describe the time-dependent characteristics of the bone remodeling process. Based on the work of Lemaire et al. [23], the last two papers [3,4] developed an extended bone-cell population model by incorporating the following significant modifications:

a rate equation describing changes in bone volume

a rate equation describing TGF-β concentration as a function of resorbed bone volume

RANKL and OPG expression on osteoblastic cells at different stages of maturation

new activator/repressor functions based on enzyme kinetics

Inspired by current advances in bone biology experiments and based on the work of Pivonka et al. [3] as well as the model presented in Wang, Qin, and Kalyanasundaram [24], Qin and Wang [2] developed a cell population dynamics model incorporating the following features, which provide deeper understanding of the mechanical bone remodeling mechanism at the cellular level:

1. The interstitial fluid shear stress rate R_{IFSS} in the lacuno-canalicular porosity structure is proposed as the physical mediator of mechanotransduction by osteocytes in mathematical formulation.

2. Three rate equations describing changes of osteocytes, NO, and PGE_2 are incorporated in this model.

3. The effect of loading frequency on bone mechanosensitivity is investigated.

4. The influence of the number of loading cycles during a loading day and the recovery of bone mechanosensitivity between two loading bouts as well as during the whole loading period are explored.

5. The Hill equation is extended to two ligands binding to the same cell.

6. A standard BFE is defined to measure the comprehensive biomechanical benefit for bone after remodeling under mechanical loading.

The use of this model for quantitative analysis has already provided an insight into the effects of mechanical loading on cortical bone remodeling [1,2]. This understanding is useful for the development of effective nonpharmacological therapies to combat bone-related pathologies. In the field of bone tissue engineering, this improved understanding of how mechanical conditions affect the formation of bone components by cells at a local level is essential for the generation of functionally appropriate tissues.

7.2 Development of Mathematical Formulation

7.2.1 RANK–RANKL–OPG Signaling Pathway

The assumption that a coupling mechanism must exist between bone formation and resorption was first articulated in 1964 [25]. However, the exact molecular mechanism that describes the interaction between cells of the osteoblastic and osteoclastic lineages was only recently identified [26]. Recent breakthroughs in our understanding of osteoclast differentiation and activation have come from the analysis of a family of biologically related TNF receptor (TNFR)/TNF-like proteins: OPG, the receptor activator of nuclear factor (NF)-κβ (RANK), and RANK ligand (RANKL), which together regulate osteoclast function [6,27].

With the discovery of the RANK–RANKL–OPG system, a revolutionary understanding of osteoclastogenesis was born. In general, the pathway involves three major components: (1) the receptor activator of nuclear factor κ-β (NF-κ-β) (RANK) is expressed on the surface of hemopoietic precursor cells (here referred to as osteoclast precursor cells); (2) RANKL, a polypeptide found on the surface of osteoblastic cells and proteolytically released as a soluble form; and (3) OPG, a "decoy receptor" molecule released by osteoblastic cells [4].

As far as the local regulation of bone cell function is concerned, after the recent discovery of the RANK–RANKL–OPG system, we have a clearer picture regarding the control of osteoclastogenesis and bone remodeling in general. The main switch for osteoclastic bone resorption is the RANKL [28], a cytokine that is released by preosteoblasts [3]. Its action on the RANK receptor is regulated by OPG, a decoy receptor, which is also derived from osteoblastic lineage-active osteoblasts [3]. Osteoclast-to-osteoblast cross talk occurs mostly through growth factors, such as TGF-β, which are released from the bone matrix during resorption. The action of TGF-β on bone cells can be accounted for by means of activator and/or repressor functions. TGF-β stimulates differentiation of uncommitted osteoblast precursor cells (OBU), but it inhibits differentiation of osteoblast precursor cells (OBP) [3]. Thus, the action of TGF-β on osteoblastic cells leads to an increase of the pool of osteoblast precursor cells, as shown in Figure 7.1. If TGF-β is removed from the system or becomes inactivated, osteoblast precursor cells can differentiate to become active osteoblastic cells (OBA). On the other hand, TGF-β has been found to promote osteoclast apoptosis [29].

The RANK–RANKL–OPG signaling pathway between osteoblasts and osteoclasts, PTH and the dual action of TGF-β is diagrammed in Figure 7.1. Given the large number of different cell types involved in active bone-cell differentiation, only a representative subset of cell types is included in the model described in this chapter. As a minimalist realistic approximation, four cell types of the osteoblast lineage (with two being state variables) and

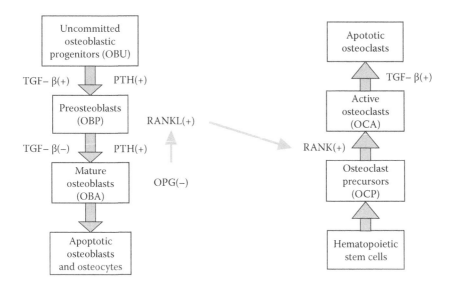

FIGURE 7.1
Illustration of bone cell model including the RANK–RANKL–OPG signaling pathway, PTH, and dual action of TGF-β. A (+) or (–) beside a factor represents stimulatory or inhibitory action by the factor.

three cell types of the osteoclast lineage (with one being a state variable) plus the osteocyte (one state variable) are considered in this chapter.

As can be seen from Figure 7.1, osteoblastic lineage derives from a large pool of mesenchymal stem cells capable of differentiating into various cells such as osteoblastic cells and adipocytes [30]. Pivonka et al. [3] denoted this pool of cells as OBU in their model. Once these cells commit to the osteoblast lineage they are generally denoted as "responding" or "preosteoblasts."

The overexpression of opposite phenotypes of OPG as well as RANKL deletion (osteopetrosis), and OPG deficiency or RANKL overexpression (osteoporosis) have led to the hypothesis that OPG and RANKL could be the mediators for the stimulatory or inhibitory effects of a variety of systemic hormones, growth factors, and cytokines on osteoclastogenesis [19]. This has recently been referred to as "the convergence hypothesis" in that the activities of the resorptive and antiresorptive agents "converge" at the level of these two mediators, whose final ratio controls the degree of osteoclast differentiation, activation, and apoptosis [31].

In addition to the RANK–RANKL–OPG signaling pathway discussed, the following two pathways are also important in bone remodeling analysis. One is the *TGF-β/nodal/activin and BMP/GDF signaling pathway*. TGF-β superfamily members are classified into the TFG-β/nodal/activin group and BMP/GDF group. TFG-β/nodal/activin signals are transduced through type I and type II receptors for each member to R-SMAD proteins, such as SMAD2 and SMAD3, while BMP/GDF signals are transduced through type I and type II receptors for each member to R-SMAD proteins, such as SMAD1, SMAD5, and SMAD8. Phosphorylated R-SMADs associated with SMAD4 are then translocated to the nucleus to activate transcription of target genes [32]. BMP/GDF family genes within the human genome have been extensively studied (http://www.gene.ucl.ac.uk, accessed March 2011); however, transcriptional regulation of BMP/GDF family members by the canonical Wnt signaling pathway remains unclear.

Another is the *Wnt/β-catenin signaling and Wnt antagonists*. Wnts constitute a family of proteins important in cell differentiation, particularly playing a critical role in OB cell differentiation and bone formation. When Wnt receptor binding interactions are absent, β-catenin is phosphorylated by glycogen-synthase kinase-3β (GSK-3β), leading to the degradation of β-catenin in the proteasome. Upon binding of Wnt to frizzled receptors and to the low-density lipoprotein receptor-related protein (LRP) coreceptors-5 and -6, the activity of GSK-3β is inhibited, leading to the stabilization of β-catenin and its translocation to the nucleus. There, it associates with T-cell factor (TCF) 4 or lymphoid enhancer binding factor (LEF) 1 to regulate gene transcription [33].

Sclerostin (SOST), produced by bone cells, has recently emerged as an important modulator of anabolic signaling pathways in bone, particularly PTH stimulation and mechanical loading. By virtue of its relatively exclusive expression in bone and its role in repressing bone formation from the extracellular space, SOST is an attractive target for anabolic therapeutics [34]. These facts make Wnt a suitable target to derive a bone anabolic response [35]. On another note, Wnt pathway components, including Wnts, Fzds, Lrps, and Tcf family members, are also expressed in osteoclast lineage cells [36]. Thus, Wnts/β-catenin signaling appears to reduce bone resorption. It remains for further investigation to resolve fully the scope of Wnt influences on bone metabolism.

7.2.2 Mechanotransduction in Bone

The great majority of the cells of bone tissue, some 95% in the adult skeleton, are osteocytes lying within the bone matrix and bone lining cells lying on the bone surface [37]. Both osteocytes and bone lining cells are terminally differentiated osteoblasts and have long been considered metabolically inactive, with limited roles in bone biology. However, osteocytes remain in contact with the bone surface cells and with neighboring osteocytes via long, slender cell processes that are connected by means of gap junctions [38,39]. Their abundance and connectivity thus make them a three-dimensional network for sensing mechanical strains.

The work reported in Nijweide, Burger, and Klein-Nulend [40] indicated that osteocytes are the professional mechanosensory cells of bone and lacuno-canalicular porosity is the structure that mediates mechanosensing. It has also become clear that dynamic mechanical load causes fluid flow in the lacuno-canalicular network [41]. Experiments in vivo [42] have indicated that this fluid flow serves as the physical mediator of the mechanotransduction of osteocytes and it is the fluid flow shear stress [43] that stimulates osteocytes within minutes to produce signaling molecules [44] such as prostaglandins (especially PGE_2) [45] and NO [46], which modulate the activities of osteoblasts and osteoclasts, thus completing the transduction from mechanical stimuli to biochemical signals [47]. NO is a strong inhibitor of bone resorption and acts by inhibiting RANKL expression in osteoblast precursors while increasing OPG production in active osteoblasts, thus decreasing the RANKL–OPG equilibrium and leading to reduced recruitment of osteoclasts and positive bone formation [48]. Alternatively, PGE_2 has strong osteogenic effects that contribute to increases in osteoblast differentiation from marrow stromal cells through the EP_4 receptor [49].

7.2.3 Mathematical Model

The schematic diagram of the mathematical model structure of bone remodeling caused by mechanical loading is shown in Figure 7.2.

In biochemistry, the Hill equation is used to describe the fraction of the macromolecule saturated by a ligand as a function of the ligand concentration; it is used in determining the degree of cooperativeness of the ligand

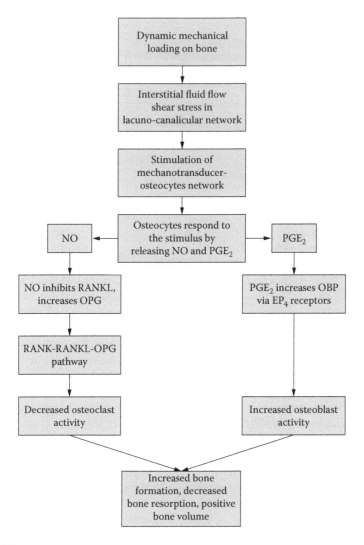

FIGURE 7.2
Schematic diagram of the mathematical model structure of cellular level bone remodeling caused by mechanical loading.

binding to the enzyme or receptor. It was originally formulated by Hill in 1910 [50] to describe the sigmoidal O_2 binding curve of hemoglobin:

$$\theta = \frac{L^n}{K_d + L^n} = \frac{L^n}{K_A^n + L^n} \tag{7.1}$$

where
  θ is the fraction of occupied sites where the ligand can bind to the active
    site of the receptor protein

L is the free ligand concentration

K_d is the apparent dissociation constant derived from the law of mass action

K_A is the ligand concentration producing half occupation that is also the microscopic dissociation constant

n is the Hill coefficient

A coefficient of $n = 1$ indicates completely independent binding, regardless of how many additional ligands are already bound. Numbers of $n > 1$ indicate positive cooperativeness, meaning that once one ligand molecule is bound to the enzyme, its affinity for other ligand molecules increases. Conversely, numbers of $n < 1$ indicate negative cooperativeness, such that once one ligand molecule is bound to the enzyme, its affinity for other ligand molecules decreases.

In cell biology, cell responses such as differentiation, proliferation, and apoptosis are all related to various ligand–receptor reactions, of which some are stimulatory and others are inhibitory [3]. The same observations are true for the production of molecules due to receptor–ligand interactions. Pivonka et al. [3] assumed that a cellular process is governed by a single factor x, which is a ligand that governs the production rate of a cell or molecule z through binding to its receptor on the cell. They then expressed the rate of production z per unit time as a function of the concentration x in its active form x^* as

$$z = f(x^*) \tag{7.2}$$

The input function f is, in general, a monotonic, S-shaped function. In modeling the cell responses governed by Equation (7.2), the Hill equation is often used to describe the molecular input function. The activation ("act" for short) and repression ("rep" for short) forms of the Hill equation [51] for the production rate of a new cell or molecule are [3]

$$f\left(x^*\right) = \beta \cdot \Pi_{act} = \frac{\beta x^*}{K_1 + x^*} \tag{7.3}$$

$$f\left(x^*\right) = \beta \cdot \Pi_{rep} = \frac{\beta}{1 + \dfrac{x^*}{K_2}} \tag{7.4}$$

where β is the maximal production rate of molecule z, Π_i ($i = act, rep$) is the input function, and K_1 and K_2 are activation and repression coefficients. Note here that we have already assumed that the Hill coefficient equals one. In the model described in this chapter, Qin and Wang [2] extended the Hill equation to the case where two ligands, x and y, both affect the production of z through binding to their respective receptors on the same cell. Then the production rate of z can be expressed as

$$z = f(x^*, y^*) = \beta \left(k_x \cdot \Pi_{act/rep}^{x^*} + k_y \cdot \Pi_{act/rep}^{y^*} \right) \tag{7.5}$$

where K_x and K_y are the relative influence of ligands x and y, respectively, as a percentage in the cellular process, and $K_x + K_y = 1$.

For convenience, in the following related formulation, the abbreviated forms are used for the factors involved. As in Figure 7.1, Qin and Wang [2] used OBU for uncommitted osteoblastic progenitors, OBP for preosteoblast, OBA for mature osteoblast, OCP for osteoclast precursor, OST for osteocyte, and OCA for active osteoclasts. RL is used for RANKL, RK for RANK, Tβ for TGF-β, and P2 for PGE$_2$; OPG, NO, and PTH remain unchanged.

The equations governing the evolution of the number of osteoblastic and osteoclastic cells in each maturation stage are simply balance equations [23], which means that each cell stage is fed by an entering flow and is emptied by the outgoing flow of differentiated or apoptotic cells (Figure 7.1). As a result, utilizing Figures 7.1 and 7.2 and based on the formulation in Pivonka et al. [3], the bone cell population dynamics can be written as follows:

$$\frac{dOBP}{dt} = D_{OBU} \cdot \left(k_{T\beta} \cdot \Pi_{act,OBU}^{T\beta} + k_{P2} \cdot \Pi_{act,OBU}^{P2} \right) - D_{OBP} \cdot OBP \cdot \Pi_{rep,OBP}^{T\beta} \tag{7.6}$$

$$\frac{dOBA}{dt} = D_{OBP} \cdot OBP \cdot \Pi_{rep,OBP}^{T\beta} - A_{OBA} \cdot OBA \tag{7.7}$$

$$\frac{dOST}{dt} = T_{OBA} \cdot OBA - A_{OST} \cdot OST \tag{7.8}$$

$$\frac{dOCA}{dt} = D_{OCP} \cdot \Pi_{act,OCP}^{RL} - A_{OCA} \cdot OCA \cdot \Pi_{act,OCA}^{T\beta} \tag{7.9}$$

where

$k_{T\beta}, k_{P2}$ are the relative influence constants of TGF-β/PGE$_2$ bindings in OBU differentiation

$\Pi_{act,OBU}^{T\beta}$ and $\Pi_{act,OBU}^{P2}$ are the activator functions related to, respectively, TGF-β and P2 bindings to their receptors on OBU

$\Pi_{rep,OBP}^{T\beta}$ and $\Pi_{act,OCA}^{T\beta}$ are the repressor/activator functions related to TGF-β binding to its receptors on OBP and OCA

$\Pi_{act,OCP}^{RANKL}$ is the activator function related to RANKL binding to its receptor RANK expressed on OCP

D_{OBU}, D_{OBP} and D_{OCP} are, respectively, differentiation rates of uncommitted osteoblast progenitors, osteoblast precursor cells, and preosteoclasts

A_{OBA} is the rate of elimination of active OBA
A_{OCA} is the rate of elimination of active OCA
A_{OST} is the apoptosis rate of active OST

All the constants and their descriptions used here can be found in Table 7.1. Equations (7.6)–(7.9) represent cell balance equations where changes of each cell population (i.e., OBP, OBA, OST, and OCA) are caused by the addition and removal of cells of the respective lineage. Addition of cells occurs naturally by proliferation and differentiation of precursor cells, and elimination of cells is caused by apoptosis (or by external extraction of cells). Differentiation and apoptosis are regulated by several activator and repressor functions—for example, TGF-β binding to its receptors on uncommitted osteoblast progenitors promoting their differentiation ($\Pi_{act, OBU}^{T\beta}$), or TGF-β binding to its receptors on osteoblast precursor cells inhibiting their differentiation ($\Pi_{rep, OBP}^{T\beta}$).

Note that in this hierarchical model there are two different time scales. A short time scale, which is less than 12 hours, is required to describe the production of NO and PGE_2, and a longer time scale of several months (up to 12 months) is needed to capture the effects of these two factors and others on bones.

Bone matrix is the largest source of TGF-β in the body [52]. Indeed, the content of TGF-β in dried bone powder is approximately 1000-fold greater than the levels required for osteoblastic stimulation [53]. TGF-β, as well as other growth factors and specific components embedded in the bone matrix, is released by osteoclasts during bone resorption [54]. The effect of TGF-β on osteoblasts is bidirectional, depending upon the state of maturation of the osteoblasts [23].

On the one hand, TGF-β has the potential to stimulate osteoblast recruitment, migration, and proliferation of osteoblast precursors (meaning OBPs in our model) [54]. On the other hand, TGF-β inhibits terminal osteoblastic differentiation into OBAs [55]. Based on the preceding discussion, the release rate of TGF-β can be evaluated as follows [3]. Considering the fact that the osteoblast lineage originates from a large population of OBU, the progenitors express a specific TGF-β receptor that, once activated, leads to differentiation of those progenitors into OBP [3]. In the model presented in Pivonka et al. [3], they assumed that the release rate of TGF-β is proportional to OCA:

$$\frac{d(T\beta)}{dt} = \alpha \cdot K_{res}OCA - \tilde{D}_{T\beta} \cdot T\beta + S_{T\beta} \qquad (7.10)$$

where $S_{T\beta}$ is a source/sink term for TGF-β, α is a proportionality constant expressing the TGF-β content stored in the bone volume, K_{res} is the relative bone resorption rate, and $\tilde{D}_{T\beta}$ is a constant degradation rate.

In the model described here, it is assumed that the release rate of TGF-β from the bone matrix is constant, and the binding of TGF-β to its receptors is

TABLE 7.1

Parameter Values and Descriptions

Symbol	Unit	Value	Description
D_{OBU}	pM/day	7×10^{-4}	Differentiation rate of uncommitted OB progenitors
D_{OBP}	pM/day	5.348	Differentiation rate of preosteoblasts
D_{OCP}	pM/day	2.1×10^{-3}	Differentiation rate of preosteoclasts
A_{OBA}	pM/day	1.890×10^{-1}	Rate of elimination of OBA
A_{OCA}	pM/day	7.000×10^{-1}	Rate of elimination of OCA
A_{OST}	pM/day	3.1×10^{-2}	Rate of elimination of OST [20,21]
$K_{D1,T\beta}$	pM	4.545×10^{-3}	Activation coefficient related to TGF-β binding on OBU
$K_{D2,T\beta}$	pM	1.416×10^{-3}	Repression coefficient related to TGF-β binding on OBP
$K_{D3,T\beta}$	pM	4.545×10^{-3}	Activation coefficient of TGF-β binding on OCA
$K_{D4,PTH}$	pM	1.500×10^{-2}	Activation coefficient for RANKL$_{eff}$ on OBP related to PTH binding
$K_{D7,PTH}$	pM	2.226×10^{-1}	Repression coefficient for OPG production related to PTH binding on OBA
$K_{D8,RL}$	pM	1.500×10^{-2}	Activation coefficient related to RANKL binding on OCP
$K_{D9,NO}$	pM	1.573×10	Activation coefficient for OPG production on OBA related to NO [20,21]
$K_{D10,NO}$	pM	2.189×10	Repression coefficient for RANKL production on OBP related to NO [20,21]
$K_{D11,P2}$	pM	3.674	Activation coefficient for OBU differentiation related to PGE$_2$ [20,21]
RK	pM	1×10	Unchanged concentration of RANK
R_{RL}	—	3×10^6	Maximum RANKL on OBP
β_{PTH}	pM/cell	2.5×10^2	Synthesis rate of systemic PTH
β_{RL}	pM/cell	1.684×10^4	Production rate of RANKL per OBP
β_{OPG}	pM/cell	1.464×10^8	Production rate of OPG per OBA
$\tilde{D}_{PTH}$	pM/day	8.6×10	Rate of degradation of PTH
$\tilde{D}_{RL}$	pM/cell	1.013×10	Rate of degradation of RANKL
$\tilde{D}_{OPG}$	pM/cell	3.5×10^{-1}	Rate of degradation of OPG
$\tilde{D}_{T\beta}$	pM/cell	1×10^0	Rate of degradation of TGF-β
$\tilde{D}_{NO}$	pM/cell	1×10^3	Rate of degradation of NO [20,21]
$\tilde{D}_{P2}$	pM/cell	1×10^2	Rate of degradation of T PGE$_2$ [20,21]
$k_{T\beta}$	—	0.5	Relative influence of TGF-β binding in OBU differentiation
k_{P2}	—	0.5	Relative influence of PGE$_2$ in OBU differentiation
k_{PTH}	—	0.7	Relative influence of PTH binding in production of OPG in OBA
k_{NO}	—	0.3	Relative influence of NO in production of OPG in OBA
$K_{A1,RL}$	pM^{-1}	1×10^{-3}	Association binding constant RANKL–OPG

TABLE 7.1 (*Continued*)

Parameter Values and Descriptions

Symbol	Unit	Value	Description
$K_{A2,RL}$	pM^{-1}	3.412×10^{-2}	Association binding constant RANKL–RANK
OPG_{max}	pM	2×10^{8}	Maximum possible OPG concentration
α	%	1	TGF-β content stored in bone matrix
K_{res}	Day^{-1}	1	Relative rate of bone resorption
K_{for}	Day^{-1}	1.571	Relative rate of bone formation
K_{to}	Day^{-1}	1.552×10	Relative rate of bone turnover [20,21]
K_{OBA}	pM/day	0.15	Rate of trapped OBA in bone matrix [20,21]
K_{NO}	pM/day	2×10^{4}	Secretion rate of NO by osteocytes [20,21]
K_{P2}	pM/day	1×10^{2}	Secretion rate of PGE$_2$ by osteocytes [20,21]
T_{acc}	Day	24	Time constant describing the rate at which accommodation takes place [56]
τ	h	6	A time constant [57]

Sources: Most parameter values are from Pivonka P. et al. *Bone* 43 (2): 249–263 (2008), and Lemaire V. et al. *Journal of Theoretical Biology* 229 (3): 293–309 (2004), except where otherwise indicated.

much faster than changes in the number of active osteoclasts. Using the short time scale and a quasisteady-state assumption in Equation (7.10), the expression of TGF-β can be given by [3]

$$T\beta = \frac{\alpha \cdot K_{res} OCA + S_{T\beta}}{\tilde{D}_{T\beta}} \tag{7.11}$$

Consequently, the activation and repression forms of TGF-β can be obtained by substituting Equation (7.11) into Equations (7.3) and (7.4) as

$$\Pi_{act,OBU}^{T\beta} = \frac{T\beta}{K_{D1, T\beta} + T\beta} \tag{7.12}$$

$$\Pi_{rep,OBP}^{T\beta} = \frac{1}{1 + T\beta / K_{D2, T\beta}} \tag{7.13}$$

$$\Pi_{act,OCA}^{T\beta} = \frac{T\beta}{K_{D3, T\beta} + T\beta} \tag{7.14}$$

Applying the law of mass action [23] used to describe the reactions of receptors and corresponding ligands, the formulations including PTH with its receptor, RANKL with OPG, and RANKL with RANK can be found based on the mathematical model presented in Chapter 6 or in Wang, Qin, and Kalyanasundaram [24]. The model presented in Pivonka et al. [3] included

PTH in the RANK–RANKL–OPG pathway model by considering the effects of PTH on bone cells as a regulator of RANKL and OPG production. They assumed that all receptor–ligand binding reactions can be written in the following form:

$$A + B_j = C_j \tag{7.15}$$

where A and B_j are two reactants and C_j represents the molecule complex formed. The index j was introduced in Pivonka et al. for cases where molecule A is involved in several (say, N) reactions such as RANKL, both of which bind to OPG and RANK ($N = 2$ in this case).

Using the principle of mass action kinetics, Pivonka et al. [3] presented the concentration balance equation for all molecules as follows:

$$\frac{dA}{dt} = -\sum_{j=1}^{N} k_{j,f} A \cdot B_j + \sum_{j=1}^{N} k_{j,r} C_j + S_A \tag{7.16}$$

$$\frac{dB_j}{dt} = -k_{j,f} A \cdot B_j + k_{j,r} C_j + S_{B_j} \tag{7.17}$$

$$\frac{dC_j}{dt} = k_{j,f} A \cdot B_j - k_{j,r} C_j \tag{7.18}$$

where $k_{j,f}$ and $k_{j,r}$ are the forward and reverse reaction rate constants, and S_A and S_{B_j} are source and sink terms, which express whether the overall contribution adds or removes mass during a chemical reaction. They are the sum of a production rate term (P_A, P_{B_j}) and a degradation term (D_A, D_{B_j}) [3]:

$$S_A = P_A + D_A \quad \text{and} \quad S_{B_j} = P_{B_j} + D_{B_j} \tag{7.19}$$

Then, Pivonka et al. assumed that binding reactions leading to upregulation and downregulation of molecules are much faster than the cell responses. As a result, the binding reactions can be assumed as quasisteady states. Using the steady-state assumption, Equations (7.16)–(7.18) yield

$$S_A = 0 \quad \text{and} \quad S_{B_j} = 0 \tag{7.20}$$

To calculate A, the production rate of A is decomposed into an endogenous term and an external "dosing term" and degradation of A is assumed to be proportional to its concentration:

$$P_A = P_{A,e}(t) + P_{A,d}(t) \tag{7.21}$$

$$D_A = -\tilde{D}_A A \tag{7.22}$$

where $\tilde{D}_A$ is a constant degradation rate of substrate A. Pivonka et al. further assumed that endogenous production is regulated by a ligand and that production cannot exceed a maximum level of concentration (A_{max}):

$$P_{A,e}(t) = \beta_A \cdot \Pi_{act/rep}^A \cdot \left(1 - \frac{A}{A_{max}}\right)$$

(7.23)

Substituting Equations (7.22) and (7.23) into Equation (7.20) gives

$$A = \frac{\beta_A \cdot \Pi_{act/rep}^A + P_{A,d}(t)}{\dfrac{\beta_A \cdot \Pi_{act/rep}^A}{A_{max}} + \tilde{D}_A}$$

(7.24)

which can be used to derive the expression of PTH. Pivonka et al. [3] took PTH as a regulator of RANKL and OPG production. The assumptions were made that PTH endogenous production was constant (i.e., βPTH = constant and $\Pi_{act/rep}^{PTH} = 1$) and $PTH_{max} \gg PTH$, and that the binding of PTH to its receptors on OBP and OBA was the same ($N = 1$), to obtain PTH concentration and its according activation and repression functions. Applying these assumptions to Equation (7.24), PTH can be obtained as [3]

$$PTH = \frac{\beta_{PTH} + P_{PTH,d}(t)}{\tilde{D}_{PTH}}$$

(7.25)

where β_{PTH} is the synthesis rate of systemic PTH, $P_{PTH,d}(t)$ represents an external PTH dosing term, and $\tilde{D}_{PTH}$ is the rate of degradation of PTH. Making use of Equation (7.25), the activator/repressor input functions defined in Equations (7.3) and (7.4) can be calculated as

$$\Pi_{act,OBP}^{PTH} = \frac{PTH}{K_{D4,PTH} + PTH} \quad \text{and} \quad \Pi_{act,OBA}^{PTH} = \frac{PTH}{K_{D5,PTH} + PTH}$$

(7.26)

$$\Pi_{rep,OBP}^{PTH} = \frac{1}{1 + PTH/K_{D6,PTH}} \quad \text{and} \quad \Pi_{rep,OBA}^{PTH} = \frac{1}{1 + PTH/K_{D7,PTH}}$$

(7.27)

where $K_{D4,PTH}/K_{D5,PTH}$ are activation coefficients for $RANKL_{eff}$ production on OBP/OBA related to PTH binding. $K_{D6,PTH}/K_{D7,PTH}$ are the repression coefficients for OPG production related to PTH binding on OBP/OBA. As a first approximation, Pivonka et al. assumed that the activation and repression coefficients for different osteoblastic cells were the same (i.e., where $K_{D4,PTH} = K_{D5,PTH}$ and $K_{D6,PTH} = K_{D7,PTH}$) and hence had the same activator/repressor function.

Qin and Wang [2] argued that NO stimulates the production of OPG expressed in OBA (see Section 7.2.2) while PTH downregulates OPG production of OBA [24]. Therefore, based on the work of Pivonka et al. [3] and using Equation (7.5), Qin and Wang obtained the expression of OPG concentration as

$$OPG = \frac{\beta_{OPG} \cdot OBA \cdot \left(k_{PTH} \cdot \Pi_{rep,OBA}^{PTH} + k_{NO} \cdot \Pi_{act,OBA}^{NO}\right) + P_{OPG,d}(t)}{\dfrac{\beta_{OPG} \cdot OBA \cdot \left(k_{PTH} \cdot \Pi_{rep,OBA}^{PTH} + k_{NO} \cdot \Pi_{act,OBA}^{NO}\right)}{OPG_{max}} + \tilde{D}_{OPG}} \tag{7.28}$$

where
β_{OPG} is the production rate of OPG per OBA
k_{PTH} is the relative influence of PTH binding in the production of OPG in OBA
k_{NO} is the relative influence of NO in the production of OPG in OBA
$P_{OPG,d}(t)$ is an external OPG administration term
$\tilde{D}_{OPG}$ is the rate of degradation of OPG
OPG_{max} is the maximum possible OPG concentration

Further, NO inhibits RANKL expression in OBP and PTH upregulates the RANKL "effective carrying capacity" of OBP [24]. Building on Pivonka et al. [3], Qin and Wang obtained the concentration of RANKL as

$$RL = \left(\frac{R_{RL} \cdot OBP \cdot \Pi_{act,OBP}^{PTH}}{1 + K_{A1,RL} \cdot OPG + K_{A2,RL} \cdot RK}\right)$$
$$\cdot \left(\frac{\beta_{RL} \cdot OBP \cdot \Pi_{rep,OBP}^{NO} + P_{RL,d}(t)}{\beta_{RL} \cdot OBP \cdot \Pi_{rep,OBP}^{NO} + \tilde{D}_{RL} \cdot R_{RL} \cdot OBP \cdot \Pi_{act,OBP}^{PTH}}\right) \tag{7.29}$$

where
R_{RL} is the maximum RANKL on OBP
$K_{A1,RL}$ is the association binding constant RANKL–OPG
$K_{A2,RL}$ is the association binding constant RANKL–RANK
β_{RL} is the production rate of RANKL per OBP
$P_{RL,d}(t)$ is an external RANKL administration term
$\tilde{D}_{RL}$ is the rate of degradation of RANKL

Then, the activation function of RANKL on differentiation of osteoclast precursor cells OCP can be obtained using Equations (7.3) and (7.29):

$$\Pi_{act,OCP}^{RL} = \frac{RL}{K_{D8,RL} + RL} \tag{7.30}$$

where $K_{D8,RL}$ is the activation coefficient related to RANKL binding on OCP.

Similarly, the unknown parameters $\Pi^{NO}_{act,OBA}$, $\Pi^{NO}_{rep,OBP}$, and $\Pi^{P2}_{act,OBU}$ can be written as [3]

$$\Pi^{NO}_{act,OBA} = \frac{NO}{K_{D9,NO} + NO} \tag{7.31}$$

$$\Pi^{NO}_{rep,OBP} = \frac{1}{1 + NO/K_{D10,NO}} \tag{7.32}$$

$$\Pi^{P2}_{act,OBU} = \frac{P2}{K_{D11,P2} + P2} \tag{7.33}$$

which relate directly to the concentrations of NO and PGE$_2$ caused by mechanical loading. Here, $K_{D9,NO}$ is the activation coefficient for OPG production on OBA related to NO, $K_{D10,NO}$ is the repression coefficient for RANKL production on OBP related to NO, and $K_{D11,P2}$ is the activation coefficient for OBU differentiation related to PGE$_2$.

Here, a loading regime is defined as Equation (7.34) (this equation is also widely used in animal tests [12,58]): The number of loading cycles during a training day is N, T_{rest} (h) is the rest time between loading bouts, and n is the number of loading bouts per day. The amplitude A(Pa) and frequency (f[Hz]) of the interstitial fluid shear stress (*IFSS*) caused by the loading can be measured using the method in Bergmann, Graichen, and Rohlmann [59], and therefore the peak fluid shear stress rate R_{IFSS} (Pa-Hz) can be defined as [46]

$$R_{IFSS} = 2\pi \cdot A \cdot f \tag{7.34}$$

In Qin and Wang [2], the interstitial fluid flow is formulated only in terms of load, as can be observed in Equation (7.34); this means that as an intermediate variable it is not actually modeled. However, it is still worth stating that the interstitial flow is the physical mediator of mechanotransduction by osteocytes. This completes the crucial transduction from mechanical stimuli to biochemical signals, which is now concluded for the first time in the field after an extended literature review. This conclusion will help other researchers in the field to understand the mechanism of mechanotransduction of bone remodeling under mechanical stimulus and to propose other possible models in the future.

To study the sensitivity of bone remodeling to mechanical loading, the mechanosensitivity of osteocytes (MS_{OST}) is defined with the frequency (f), number of loads per day (N), the rest time between bouts (T_{rest}), and the length of loading period (t). Experimental results indicate that loading has no effect on bone formation if its frequency is less than 0.5 Hz [60], and the sensitivity of bone changes little when the loading frequency exceeds 10 Hz [61]

However, Rubin, Xu, and Judex [62] demonstrated that loading at frequencies up to 90 Hz can still have an effect on bone formation. A logarithmic function is used to describe the relationship between frequency (f) and osteocyte mechanosensitivity (MS_{OST}):

$$MS_{OST} \propto \ln(f+0.5), \, MS_{OST} \geq 0 \tag{7.35}$$

Data from Burr, Robling, and Turner [63] show that osteocyte sensitivity to mechanical loading is proportional to $1/(N+1)$:

$$MS_{OST} \propto \frac{1}{N+1} \tag{7.36}$$

This means that bone loses more than 95% of its mechanosensitivity after only 20 loading cycles. It can be imagined that osteocytes will regain their sensitivity after a period of rest between loading bouts. Robling, Burr, and Turner [57] demonstrated the following relationship:

$$MS_{OST} \propto \left(2 - e^{-T_{rest}/\tau}\right) \tag{7.37}$$

where τ is a time constant equal approximately to 6 h [57]. It is noted that 98% of bone mechanosensitivity is regained after 24 h of rest.

Bone cells accommodate to routine loading, which means that bone mechanosensitivity drops as the loading period extends [58]. Qin and Wang [2] hypothesized that bone mechanosensitivity follows the relationship with loading period t (days):

$$MS_{OST} \propto e^{-t/T_{acc}} \tag{7.38}$$

where T_{acc} is the time constant describing the rate at which accommodation takes place—here assumed to be 24 days. Making use of Equations (7.35)–(7.38), osteocyte mechanosensitivity can be written as

$$MS_{OST} = K_{MS} \cdot \frac{\ln\left(f+0.5\right)}{N+1} \cdot \left(2 - e^{-T_{rest}/\tau}\right) \cdot e^{-t/T_{acc}} \tag{7.39}$$

where K_{MS} is a proportionality constant.

In most animal experiments, the mechanical stimulation is applied for no more than 1 h per day and lasts for several months [12,64]. The conclusions from animal studies are that limited benefit is derived from additional loading cycles above approximately 40 cycles per day [6], and it has been clear that NO and PGE_2 production appear within minutes [44] when the mechanical loading starts, ending several hours after the loading stops [65]. In this chapter, a short time scale is used to describe NO and PGE_2 production caused by a mechanical stimulus, which is assumed to be much faster than changes in the number of osteocytes (long time scale) in remodeling BMUs. Using Equations (7.34) and (7.39), and based on experimental results [46,66], the concentration changes of NO and PGE_2 during the bone remodeling process are defined as

$$\frac{dNO}{dt} = K_{NO} \cdot R_{IFSS} \cdot OST \cdot n \int_0^N MS_{OST}\, dN - \tilde{D}_{NO} \cdot NO \tag{7.40}$$

$$\frac{dP2}{dt} = K_{P2} \cdot R_{IFSS} \cdot OST \cdot n \int_0^N MS_{OST}\, dN - \tilde{D}_{P2} \cdot P2 \tag{7.41}$$

where
$\quad K_{NO}$ is the secretion rate of NO by osteocytes
$\quad n$ is the number of loading bouts per day
$\quad K_{P2}$ is the secretion rate of PGE_2 by osteocytes
$\quad \tilde{D}_{NO}$ is the rate of degradation of NO
$\quad \tilde{D}_{P2}$ is the rate of degradation of T PGE_2

There are now six unknown variables—*OBP, OBA, OST, OCA, NO,* and *P2*—and six independent equations: (7.6)–(7.9), (7.40), and (7.41). This ordinary differential equation system can be numerically solved and the numerical results for each variable can be obtained.

Then, following the method used in the work of Wang et al. [24], Qin and Wang [2] assumed that bone formation and resorption rates are proportional to the number of active bone cells—that is,

$$\frac{dBMC}{dt} = K_{for} \cdot \left[OBA(t) - OBA(t_0) \right] - K_{res} \cdot \left[OCA(t) - OCA(t_0) \right] \tag{7.42}$$

Note that *BMC* is the bone mineral content in percentage and K_{for} and K_{res} are the relative bone formation and resorption rates. The simulation starts from a so-called steady state in which the values of model variables remain constant as initial values such as $BMC(t) = 100\%$, $OBA(t) = OBA(0)$ and $OCA(t) = OCA(0)$. Therefore, the model equations (7.6)–(7.9) become

$$D_{OBU} \cdot \Pi_{act,OBU}^{T\beta} - D_{OBP} \cdot OBP(0) \cdot \Pi_{rep,OBP}^{T\beta} = 0 \tag{7.43}$$

$$D_{OBP} \cdot OBP(0) \cdot \Pi_{rep,OBP}^{T\beta} - A_{OBA} \cdot OBA(0) = 0 \tag{7.44}$$

$$T_{OBA} \cdot OBA(0) - A_{OST} \cdot OST(0) = 0 \tag{7.45}$$

$$D_{OCP} \cdot \Pi_{act,OCP}^{RL} - A_{OCA} \cdot OCA(0) \cdot \Pi_{act,OCA}^{T\beta} = 0 \tag{7.46}$$

By solving Equations (7.43)–(7.46), the initial values of the model variables in Table 7.2 can be obtained.

TABLE 7.2

Initial Values of the Model Variables

Symbol	Unit	Value
OBA(0)	pM	0.2126991130e-3
OBP(0)	pM	0.8986869185e-5
OCA(0)	pM	0.2769166993e-3
OST(0)	pM	0.1029189256e-2

A mechanical stimulus not only can increase bone mass but also can improve bone strength by influencing collagen alignment as new bone is being formed by osteoblasts during bone turnover. A cortical bone tissue located in regions subject to predominantly tensile stresses has a higher percentage of collagen fibers aligned along the long axis of the bone [6]. In regions of predominantly compressive stresses, collagen fibers are more likely to be aligned transverse to the long axis [67]. Evidently, this arrangement functionally improves bone tensile properties with more collagen fibers oriented in the longitudinal direction [68], whereas bone compressive properties are improved by transversely oriented collagens [69].

In an experiment by Robling et al. [12], cyclic mechanical loads were applied axially along the ulna of adult rats for 16 weeks. The results indicated that the pattern of bone formation caused by loading resembled the stress distribution, with more bone formation where the stresses were highest. Robling et al. tested the mechanical properties and BMC of the sample before and immediately after the experiment stopped. They compared and analyzed the data (before and after the experiment) and found that the bone structure had improved, with a 69% increase in the second moment of area. Bone strength increased by 64% and the energy absorbed before fracture (BFE), which is of greater practical interest to clinical practice, increased by 94% while the BMC improved by only 7%.

The results demonstrate that BMC is only one of the factors that contribute to bone strength, and it is the BFE that can evaluate the effect of mechanical bone remodeling comprehensively. Thus, the use of BMC to characterize bone remodeling, as done in existing mathematical models [3,23], is not suitable to measure the effect of bone remodeling under mechanical stimulus. BMC is therefore replaced by BFE in this model.

In references 3 and 23, the BFE is used as a more appropriate standard to assess the significance of bone remodeling under mechanical stimulus. The BFE can be measured through experimentation; here, using the cell population dynamics model, it is proposed that the formulation for calculating BFE is

$$\frac{dBFE}{dt} = K_{for} \cdot \left[OBA(t) - OBA(t_0) \right] - K_{res} \cdot \left[OCA(t) - OCA(t_0) \right]$$

$$+ K_{to} \cdot \sqrt{OBA(t) + OCA(t)} \tag{7.47}$$

where K_{to} is the relative rate of bone turnover. By substituting Equation (7.42) into Equation (7.47), it can be observed that BMC contributes to BFE:

$$\frac{dBFE}{dt} = \frac{dBMC}{dt} + K_{to} \cdot \sqrt{OBA(t) + OCA(t)} \qquad (7.48)$$

Qin and Wang [2] proposed that BMC linearly contributes to BFE, which is the first term in Equation (7.48). The second term in Equation (7.48) represents the contribution of bone turnover $(OBA(t)+OCA(t))$ to BFE, in order to consider the structural effects of optimized collagen alignment on bone strength in the newly formed bone. It has been observed in experiments [6,67,69] that the contribution of bone turnover to the increment of bone strength diminishes as bone turnover increases. Qin and Wang postulated that the square root of bone turnover contributes to the derivative of BFE with respect to time. By applying the superposition principle, the relationship of BFE with BMC and bone turnover can be obtained in the form of Equation (7.48). The validity of the proposed equations is tested in the following section. This equation might not be completely accurate, but it fits the model well, at least providing an alternative understanding of the relationship between BFE, BMC, and bone turnover.

In the work reported in Qin and Wang [2], all the parameter values used in the model are from experiments, and they are not amended or modified to fit the experimental data. A cursory examination of the parameters indicates two classes. The first class corresponds to the physicochemical parameters: $K_{D1,T\beta}$, $K_{D2,T\beta}$, $K_{D3,T\beta}$, $K_{D4,PTH}$, $K_{D5,PTH}$, $K_{D6,RL}$, $K_{D7,NO}$, $K_{D8,NQ}$, $K_{D9,P2}$, T_{acc}, $K_{A1,RL}$, $K_{A2,RL}$, k_{NO}, k_{PTH}, k_{P2}, $k_{T\beta}$, β_{PTH}, β_{RL}, β_{OPG}, D_{PTH}, D_{RL}, $\tilde{D}_{OPG}$, $D_{T\beta}$, D_{NO}, $\tilde{D}_{P2}$ and τ. These parameters generally remain fixed under different physiological conditions. They are easily measured through experiments and have values reported in the literature.

The second class of parameter value may fluctuate slightly if the larger physiological environment changes. They are A_{OBA}, D_{OBU}, D_{OBP}, T_{OBA}, A_{OST}, D_{OCP}, A_{OCA}, K_{NO}, K_{P2}, K_{for}, K_{res}, K_{to}, RK, α, OPG_{max}, and R_{RL}. The values for these parameters are averaged from a range of acceptable values for each by checking the literature. In the analysis presented in Qin and Wang [2], most parameter values are referred from previous models in references 3 and 23. Values of the new parameter used in this chapter are from references 20, 21, 56, and 57 (see Table 7.1).

7.3 Numerical Investigation

Mathematical models of biology are a form of complex hypothesis. To test the validity of the hypothesis, external data are utilized to see if the model matches the experiments. Ideally, the model should be tested by as many experiments as possible to see if it is valid in a statistical sense. The reality, however, is that due to the complexity of such experiments and suitability of the specific model, there are few experiments available for comparison. In this section, the results presented in Wang [1] and Qin and Wang [2] for simulating five experiments of two different types in Saxon et al. [58] and Robling et al. [12] are briefly reviewed.

First, Qin and Wang [2] simulated the three experiments conducted in Saxon et al. [58], in which 57 female Sprague–Dawley rats were randomized to three groups: group I loading was applied for 5 weeks followed by 10 weeks of time off (1 × 5); group II loading was applied for 5 weeks followed by 5 weeks of time off and loading again for 5 weeks (2 × 5); group III loading was applied continuously for 15 weeks (3 × 5). An axial load was applied to the right ulna for 360 cycles per day, at 2 Hz, 3 days per week at 15 N. Qin and Wang simulated the effects of the three different loading schemes on BFE and compared them with the experimental results, which are shown in Figure 7.3.

The trends of BMC of loading schemes 1 × 5 and 3 × 5 are shown in Figure 7.3(a). From the dotted curve, we can see that for the 1 × 5 loading scheme, the BMC retains its initial value for the first 4 ~ 5 days of loading, which is in agreement with bone remodeling theory, and then it begins to increase almost linearly under the mechanical stimulus until the loading stops on the 35th day. After that, because of the accumulated NO and PGE_2, the BMC continues to increase but the BFR (bone formation rate) drops until BMC reaches its peak value when BFR becomes zero on the 50th day; the bone remodeling then maintains its new equilibrium and the BMC remains unchanged.

During the 3 × 5 loading scheme, the BMC follows the same pattern as in the 1 × 5 scheme for the first 5 weeks. The mechanical stimulus continues for another 10 weeks, when the BFR decreases slowly toward the end of the experiment (15 weeks' loading). By the end of the experiment, each rat's BMC was measured and compared with that before the experiment. The measured BMC increases (in percentage) are shown in the graph by the small circles and squares with the word "Experiment [58]" and are 7.6% and 10.4% for the 1 × 5 and 3 × 5 schemes, respectively—closely matching the simulation results that are shown by the small circles and squares with the words "Present model" predicting 5.7% and 11.1% increases for the 1 × 5 and 3 × 5 schemes, respectively.

The pattern of BMC with respect to time (days) for the 2 × 5 loading scheme is plotted in Figure 7.3(b). From day 0 to day 70 (5 weeks' loading plus 5 weeks' rest), the BMC demonstrates exactly the same pattern as in the 1 × 5 scheme. After the 5 weeks' break, the bone cells regain some mechanosensitivity and start to respond to the mechanical loading from day 71 as predicted, but the graph shows clearly that the BFR is less than that in the first 5 weeks, implying that the bone mechanosensitivity has not recovered fully after 5 weeks' rest. Eventually, the experiment shows a 9.8% increase, which compares well with the 7.4% from our simulation results (see the small squares with words "Experiment [58]" and "Present model" representing experimental and current simulation results).

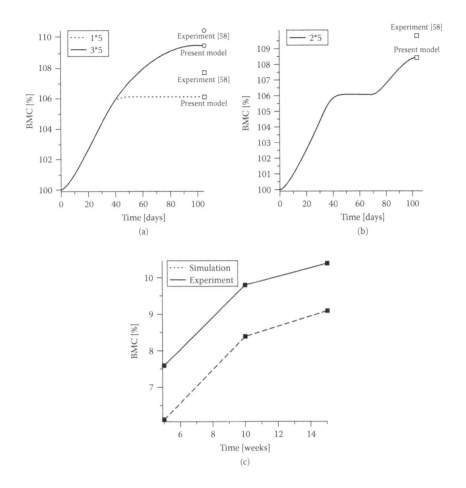

FIGURE 7.3
Simulation results of experiment in Saxon et al. [58]. (a): BMC (percentage) dynamics during the loading period 15 weeks for 1 × 5 and 3 × 5 loading scheme. (b): BMC (percentage) dynamics during the loading period 15 weeks for 2 × 5 loading scheme. (c): Trend comparison of experiment and simulation results. In (a) and (b), the small squares and circles with the words "Experiment [58]" and "Present model" represent the experimental and current simulation results, respectively, on the 105th day.

In Figure 7.3(c), the three experimental results from Saxon et al. [58] are plotted and connected with solid line segments, and the three corresponding simulation results are plotted and linked with dashed line segments. In a model of this complexity, a close match of quantitative data is mostly based on the careful choice of parameters. However, what is more important is the ability of the model to predict the qualitative pattern of the response. As can be seen from the graph, the trend of simulation is inconsistent with the experiments. For the same reason, the focus in the following comparisons is on the trends of changes (percentage of increase or decrease) rather than the values themselves. In this particular experiment, one clear conclusion is that inserting the rest period into the exercise regime does not result in a proportional decrease in bone gain. On the contrary, the bone gain in the 2 × 5 and 3 × 5 loading schemes is similar: 8.4% versus 9.1% (only 8.3% difference) in the simulation and 9.8% versus 10.4% (only 6.1% difference) in the experiment. It is much higher than the 1 × 5 loading scheme, which is 6.1% in the simulation and 7.6% in the experiment.

In another two experiments [12], the right ulnas of 26 adult female rats were subjected to 360 load cycles/day, delivered in a haversine waveform at 17 N peak force, 2 Hz, 3 days/week for 16 weeks. Half of the rats (13) were administered all 360 daily cycles in a single uninterrupted bout (360 × 1); the other half were administered 90 cycles four times per day (90 × 4), with 3 h rest time between bouts. At the end of the intervention, the BMC and BFE were measured for each rat, statistically analyzed, and compared with the nonloaded baseline control group. The simulation results and experimental results presented in Qin and Wang [2] are plotted in Figures 7.4 and 7.5.

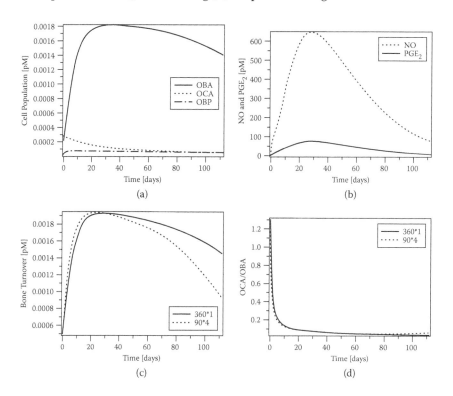

FIGURE 7.4
Simulation results of the right ulnas of 26 adult female rats subjected to 360 load cycles/day. (a): OBA, OCA, and OBP cell population dynamics during the loading period (360 × 1). (b): Messengers NO and PGE$_2$ population dynamics during the loading period (360 × 1). (c): Bone turnover (OBA + OCA) dynamics. (d): OCA/OBA during the loading period.

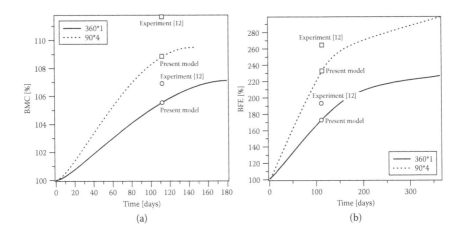

FIGURE 7.5
Simulation results and comparison with those from experiment in Robling et al. [12]. (a): BMC (percentage) dynamics during extended loading period of 170 days. (b): Energy (percentage) absorbed before bone fracture dynamics during the extended loading period of 365 days. Note that in (a) and (b) the small squares and circles with the words "Experiment [12]" and "Present model" represent the experimental and current simulation results, respectively, on the 112th day.

Using the preceding model, in addition to BMC, which can be measured easily through experiment, we can also gain insight into the cellular dynamics of OBA, OCA, and OBP during the loading period, which is difficult or perhaps impossible to measure in an experiment. Figure 7.4(a) shows the population dynamics of OBA, OCA, and OBP during the 360 × 1 loading period. It can be seen that OBA reacts quickly to the mechanical loading, with a large increase rate because of the positive effect of PGE_2 in the first 30 days. It then keeps dropping slowly to the end of the experiment, which is consistent with the decrease of PGE_2 in Figure 7.4(b). Due to the inhibitory effect of NO on OCA through the RANK–RANKL–OPG pathway, the number of OCA drops slightly as a result of limited production and degradation. Under the stimulatory effect of TGF-β (limited because of the limited activity of OCA) and PGE_2, the OBP numbers increase slightly compared with the initial value and reach almost the same value as the OCA from the 60th day.

Figure 7.4(b) depicts the population dynamics of NO and PGE_2 in the 360 × 1 loading period. NO responds to the mechanical stimulus quickly with a rapid increase in number, and PGE_2 is also found to increase, consistent with the experimental results (see Section 7.2.2). The numbers of both NO and PGE_2 start to decrease from the 30th day, probably because the bone cells accommodate to the routine loading and become desensitized.

The dynamics of bone turnover (OBA + OCA) during the loading period for the 360 × 1 and 90 × 4 loading schemes can be seen from Figure 7.4(c). From the beginning to about the 30th day, both loading schemes cause bone turnover to increase quickly at a similar rate (with the value for 90×4 slightly greater than that for 360×1). After that, it is surprising that the bone turnover in the 90×4 scheme drops more quickly than that in the 360×1 loading scheme, implying less bone cell activity in samples from the 90×4 loading scheme after the bone cells become desensitized from the 30th day.

Figure 7.4(d) presents the ratio OCA/OBA dynamics in the loading period for the 360×1 and 90×4 loading schemes, starting from $OCA/OBA \approx 1.25$ in healthy adults, which is consistent with Lemaire et al. [23]. Both ratios show almost the same pattern and value: They drop quickly once the mechanical loading applies, from 1.25 to about 0.1 on the 20th day, and then they remain almost unchanged until the end of the experiment, predicting an osteogenic outcome after loading.

Figure 7.5(a) shows the BMC dynamics during an extended loading period (more than 16 weeks) for the 360×1 and 90×4 loading schemes. The BMC remains almost at its initial value in the first week, which agrees with bone remodeling cycle theory. Then, the BFR and BMC values for the loading scheme 90×4 exceed those of the 360×1 loading scheme, which is consistent with the experimental observation [12].

As we can see from the graph, BFR in both cases diminishes as time elapses. We continue to draw the graph (which means the loading continues in the experiment) and it is interesting to see that it takes less time for the BMC to achieve peak value in the 90×4 loading scheme (150 days) than in the 360×1 loading scheme (180 days). Unfortunately, there is no experiment to verify this finding. By the end of the experiment (that is, 16 weeks' loading), our simulation shows 5.5% and 8.8% increase in BMC for the 360×1 and 90×4 loading schemes, respectively (see the small circles and squares with the words "Present model" on the solid and dotted curves). This compares well with the measured changes in experiments, in which the increases were by 6.9% and 11.7%, respectively [12] (see the small circles and squares with words "Experiment [12]").

This conclusion is significant to mechanical stimulus therapies, in that separating loading into short bouts such as 90×4 in this experiment not only ultimately achieves greater BMC but also uses less time compared with one long loading bout, such as 360×1 in this case.

Using our newly proposed standard of bone fracture energy, we calculate the dynamics of BFE in the extended loading period (that is 365 days) in Figure 7.5(b). The results show that the BFE increases almost linearly with respect to time (days) until about the 150th day in the 90×4 loading scheme and the 180th day in the 360×1 loading scheme, which matches the timing when BMC peaks in Figure 7.5(a). Then, the BFE continues to grow linearly at a slower rate, probably because the increase of BMC ceases in both cases. The distinguishing difference between Figure 7.5(a) and 7.5(b) is that the BFE continues to grow even after the BMC stops increasing, meaning that bone strength continues to benefit from the mechanical stimulus although there is hardly any gain in bone mass. Our simulation results indicate 72.8% and

131.6% increases of BFE for the 360 × 1 and 90 × 4 loading schemes, respectively, after a 16 week loading (see the small circles and squares with words "Present model" on the solid and dotted curves). This closely resembles the measured changes in the experiment, in which the BFE increased by 94% and 165%, respectively [12] (see the small circles and squares with words "Experiment [12]"), in Figure 7.5.

Comparing the increases of BMC in Figure 7.5(a) and BFE in Figure 7.5(b), the present study shows that these small gains in BMC (around 10%) impart very large increases in BFE (72%–131%) because the new bone formation is localized to the most mechanically needed sites. Consequently, it is possible to enhance fracture resistance significantly through mechanical loading such as exercises, even though most exercise intervention studies yield increases in BMC of only a few percent at most. There might be a significant difference between pharmacologically induced bone formation and loading-induced bone formation. For example, intermittent administration of PTH adds new bone mainly to the endocortical and trabecular surfaces, which makes relatively little contribution to resistance to bending [70]. Mechanical loading, on the other hand, appears to be able not only to increase bone mass but also, more importantly, to optimize the new bone formation spatially to obtain maximal bone strength.

7.4 Parametric Study of the Control Mechanism

In this section, results in Wang and Qin [71] are presented and discussed. In the analysis, perturbations are applied to the mechanical bone remodeling system in a steady state, by down- and upregulating its six differentiation and apoptosis rate parameters DF_{obu}, DF_{ocp}, DF_{obp}, A_{oba}, A_{oca}, and A_{ost}. In this case, there are six different parameters and each parameter may be either up-or downregulated. Using a simple combination theory, the number of permutations is calculated as

$$728 = \sum_{i=1}^{6} C_6^i \cdot 2^i$$

To investigate the system behavior for a wide range of changes, the exponentially changed factor is applied (1.5^{ex}) to each of the six differentiation and apoptosis rate parameters, whereby the exponent ex ranges from –10 to 10 in step increases of 0.5. The assessment of the contribution of each of the parameter combinations to the system's behavior is chosen as the responses of BMC and BFE, which are sampled on the 100th day, to stand for the maximum change. After analysis of all the combinations of 728 permutations in six model parameters, a small number of parameter combinations were identified

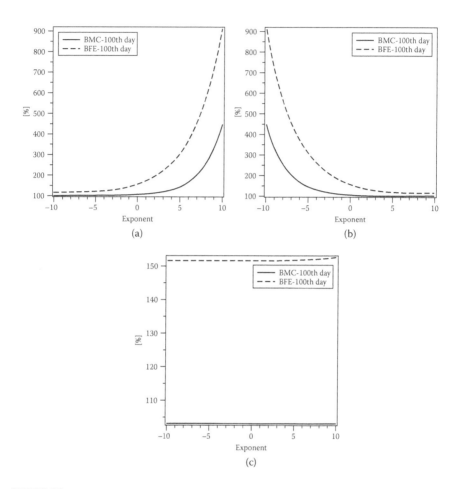

FIGURE 7.6
Physiologically unrealistic changes of BMC and BFE versus combined changes of model
parameter [1.5^{-10}–1.5^{+10}] p: (a) exponential bone growth, (b) exponential bone decrease, (c) slight
changes of bone. (p is the parameter value.)

that can lead to physiologically realistic responses, which are similar to
theoretically idealized physiological responses and are plotted in Figure 7.6.

Figure 7.6(a) and 7.6(b) shows an exponential increase and decrease of BMC
and BFE, respectively, when increasing the model parameter exponentially
(exponent *ex* from –10 to 10). This type of behavior is considered physiologically unrealistic from a biological perspective and is obtained for quite a
large range of model parameter combinations. Conversely, Figure 7.6(c) represents the other extreme, whereby only minor changes of BMC and BFE
occur during the entire range of parameter variation. These three types of
response curves are excluded from further analysis because they do not
provide an effective control mechanism for BMC and BFE.

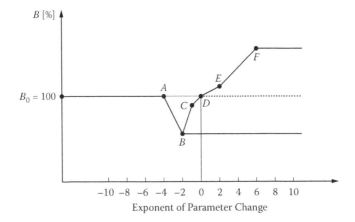

FIGURE 7.7
Schematic illustration of ideal response curve for combined changes of model parameters. (Modified from Pivonka, P. et al. *Bone* 43 (2): 249–263, 2008.)

In the work reported in Pivonka et al. [3], the "idealized" regulatory response by functionally active BMUs is discussed. As was stated earlier, the bone remodeling system starts from a steady state in which it can be identified that $\Delta BMC = 0$, $\Delta BFE = 0$, and concentrations of various hormones, growth factors that cause fluctuations of initial values of differentiation and apoptosis rates in BMUs. In order to respond to minor changes in concentrations, it is expected that BMUs should be rather insensitive to the fluctuations mentioned.

From Figure 7.7, it can be recognized that point A is the threshold concentration, which means that any change in a model parameter below A causes no change in BMC (/BFE). Further, regions around the usual operation status of BMUs should be found with relatively small gradients of change in BMC (/BFE) in response to changes in differentiation rates (regions C–D and D–E in Figure 7.7), and regions with larger gradients of change for larger changes in differentiation rates (region E–F in Figure 7.7). It is expected that this response in BMC (/BFE) changes will remain limited if the differentiation rates increase significantly (region beyond point F in Figure 7.7) because an unlimited increase of BMC (/BFE) is not physiologically realistic.

Conversely, it is expected that the rate of BMC (/BFE) change will also decrease in a limited manner if the differentiation rates decrease significantly. In fact, physiologically, it is reasonable for the BMC (/BFE) change to be zero for extremely small differentiation rates. Additionally, it can be observed from Figure 7.7 that point F marks the maximum change in BMC (/BFE) (ΔB max). Since we have a point A that is the maximum concentration that does not lead to further modifications of BMC (/BFE), there must be a transition region from point C to A that is characterized by point B, the lowest value of BMC (/BFE).

Having found a potential "ideal response curve," we can begin a search for response curves that might meet these requirements. Encouragingly, this research is able to identify a small number of curves that possess similarity to the idealized response curve. Table 7.3 summarizes all the parameter combinations that produce idealized response curves. Figure 7.8 illustrates the

TABLE 7.3

Summary of Parameter Combinations That Lead to Controlled Remodeling Processes

Number of Parameters in a Combination	Combinations of Differentiation and Apoptosis rates	Variation of Each Parameter
1	A_{OBA}	–
1	A_{OST}	–
2	D_{OBP}/A_{OBA}	+/–
2	D_{OBP}/A_{OST}	–/–
2	D_{OCP}/A_{OBA}	–/–
2	D_{OCP}/A_{OST}	–/–
2	A_{OBA}/A_{OCA}	–/+
2	A_{OCA}/A_{OST}	+/–
3	$D_{OBP}/D_{OCP}/A_{OBA}$	–/–/–
3	$D_{OBP}/D_{OCP}/A_{OBA}$	–/+/–
3	$D_{OBP}/A_{OBA}/A_{OCA}$	–/–/–
3	$D_{OBP}/A_{OBA}/A_{OCA}$	–/–/+
3	$D_{OBP}/A_{OBA}/A_{OST}$	–/–/–
3	$D_{OBP}/A_{OBA}/A_{OST}$	+/–/+
3	$A_{OBA}/A_{OCA}/A_{OST}$	–/+/+
4	$D_{OBU}/D_{OBP}/A_{OBA}/A_{OST}$	+/+/+/–
4	$D_{OBP}/D_{OCP}/A_{OBA}/A_{OCA}$	–/–/–/–
4	$D_{OBP}/D_{OCP}/A_{OBA}/A_{OCA}$	–/–/–/+
4	$D_{OBU}/D_{OCP}/A_{OBA}/A_{OST}$	–/+/–/–
4	$D_{OBP}/A_{OBA}/A_{OCA}/A_{OST}$	–/–/+/–
4	$D_{OBP}/A_{OBA}/A_{OCA}/A_{OST}$	+/–/–/+
5	$D_{OBU}/D_{OBP}/D_{OCP}/A_{OBA}/A_{OST}$	–/–/+/–/–
5	$D_{OBU}/D_{OBP}/D_{OCP}/A_{OBA}/A_{OST}$	+/–/–/+/–
5	$D_{OBU}/D_{OBP}/A_{OBA}/A_{OCA}/A_{OST}$	–/+/–/–/–
5	$D_{OBU}/D_{OBP}/A_{OBA}/A_{OCA}/A_{OST}$	+/–/+/+/–
5	$D_{OBP}/D_{OCP}/A_{OBA}/A_{OCA}/A_{OST}$	–/+/–/–/–
5	$D_{OBP}/D_{OCP}/A_{OBA}/A_{OCA}/A_{OST}$	–/–/–/+/–
5	$D_{OBP}/D_{OCP}/A_{OBA}/A_{OCA}/A_{OST}$	–/–/–/–/+
6	$D_{OBU}/D_{OBP}/D_{OCP}/A_{OBA}/A_{OCA}/A_{OST}$	+/–/–/+/–/–
6	$D_{OBU}/D_{OBP}/D_{OCP}/A_{OBA}/A_{OCA}/A_{OST}$	+/–/–/+/+/–

Note: The variation with "+" represents a parameter increase; "–" represents a parameter decrease.

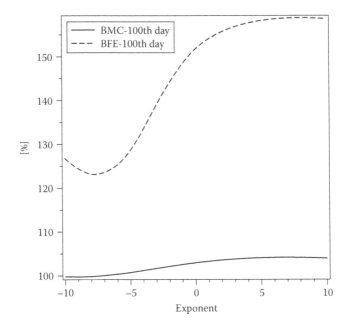

FIGURE 7.8
Typical physiologically realistic fluctuations of BMC and BFE with combinations of parameter change.

physiologically realistic response curve, which corresponds to the parameter permutation involving three parameters (A_{OBA}, A_{OCA}, and $A_{OST} = -/+/+$) and is similar to the idealized response curve shown in Figure 7.7.

It is noticed that in a bone remodeling system without consideration of mechanical stimulus, the response involving three parameters ($D_{OBU}/D_{OBP}/A_{OCA} = +/-/+$) coincides with the known physiological action of TGF-β on bone cells. TGF-β promotes differentiation of osteoblast progenitors and inhibits differentiation of osteoblast precursor cells while promoting osteoclast apoptosis [3]. However, in the case of mechanical bone remodeling, this combination causes exponential increases for both BMC and BFE similar to those illustrated in Figure 7.6(a). In other words, with the introduction of mechanical stimuli, the bone remodeling system changes. This warrants attention from biologists and other researchers.

References

1. Wang Y. N. Mathematical modeling of bone remodeling under mechanical, electromagnetic fields at the cellular level. PhD thesis, Australian National University (2010).

2. Qin Q. H., Wang Y. N. A mathematical model of cortical bone remodeling at cellular level under mechanical stimulus. *Acta Mechanica Sinica* (in press) (2012).

3. Pivonka P., Zimak J., Smith D. W., Gardiner B. S., Dunstan C. R., Sims N. A., Martin J. T., Mundy G. R. Model structure and control of bone remodeling: A theoretical study. *Bone* 43 (2): 249–263 (2008).

4. Pivonka P, Zimak J, Smith D. W., Gardiner B. S., Dunstan C. R., Sims N. A., Martin J. T., Mundy G. R. Theoretical investigation of the role of the RANK–RANKL–OPG system in bone remodeling. *Journal of Theoretical Biology* 262 (2): 306–316 (2010).

5. Frost H. M. *Intermediary organization of the skeleton*. Boca Raton, FL: CRC Press (1986).

6. Robling A. G., Castillo A. B., Turner C. H. Biomechanical and molecular regulation of bone remodeling. *Annual Review of Biomedical Engineering* 8:455–498 (2006).

7. Zerwekh J. E., Ruml L. A., Gottschalk F., Pak C. Y. C. The effects of twelve weeks of bed rest on bone histology, biochemical markers of bone turnover, and calcium homeostasis in eleven normal subjects. *Journal of Bone Mineral Research* 13 (10): 1594–1601 (1998).

8. Schneider V., Oganov V., LeBlanc A., Rakmonov A., Taggart L., Bakulin A., Huntoon C., Grigoriev A., Varonin L. Bone and body mass changes during space flight. *Acta Astronautica* 36 (8–12): 463–466 (1995).

9. Warden S. J. Breaking the rules for bone adaptation to mechanical loading. *Journal of Applied Physiology* 100 (5): 1441–1442 (2006).

10. Ducher G., Jaffre C., Arlettaz A., Benhamou C. L., Courteix D. Effects of long-term tennis playing on the muscle–bone relationship in the dominant and non-dominant forearms. *Canadian Journal of Applied Physiology* (*Revue Canadienne de Physiologie Appliqué*) 30 (1): 3–17 (2005).

11. Bass S. L., Saxon L., Daly R. M., Turner C. H., Robling A. G., Seeman E., Stuckey S. The effect of mechanical loading on the size and shape of bone in pre-, peri-, and postpubertal girls: A study in tennis players. *Journal of Bone Mineral Research* 17 (12): 2274–2280 (2002).

12. Robling A. G., Hinant F. M., Burr D. B., Turner C. H. Improved bone structure and strength after long-term mechanical loading is greatest if loading is separated into short bouts. *Journal of Bone Mineral Research* 17 (8): 1545–1554 (2002).

13. MacKay C., Lorincz C., Zernicke R. Mechanisms of bone remodeling during weight-bearing exercise. *Applied Physiology, Nutrition, and Metabolism* 31: 655–660 (2006).

14. Mullender M. G., Huiskes R., Weinans H. A physiological approach to the simulation of bone remodeling as a self organization control process. *Journal of Biomechanics* 27: 1389–1394 (1994).

15. Ruimerman R., Huiskes R., Van Lenthe G. H., Janssen J. D. A computer-simulation model relating bone-cell metabolism to mechanical adaptation of trabecular architecture. *Computer Methods in Biomechanics and Biomedical Engineering* 4 (5): 433–448 (2001).

16. Ruimerman R., Hilbers P., van Rietbergen B., Huiskes R. A theoretical framework for strain-related trabecular bone maintenance and adaptation. *Journal of Biomechanics* 38 (4): 931–941 (2005).

17. Weinans H., Huiskes R., Grootenboer H. J. The behavior of adaptive bone-remodeling simulation models. *Journal of Biomechanics* 25 (12): 1425–1441 (1992).

18. Li J., Li H., Shi L., Fok A. S. L., Ucer C., Deulin H., Horner K., Silikas N. A mathematical model for simulating the bone remodeling process under mechanical stimulus. *Dental Materials* 23 (9): 1073–1078 (2007).
19. Hadjidakis D. J., Androulakis I. I. Bone remodeling. *Annals of the New York Academy of Sciences* 1092 (1): 385–396 (2006).
20. Maldonado S., Borchers S., Findeisen R., Allgower F. Mathematical modeling and analysis of force induced bone growth. *Conference Proceedings of IEEE Engineering and Medical Biology Society* 1: 3154–3157 (2006).
21. Maldonado S., Findeisen R., Allgower F. Describing force-induced bone growth and adaptation by a mathematical model. *Journal of Musculoskeletal and Neuronal Interactions* 8 (1): 15–17 (2008).
22. Tanaka E., Yamamoto S., Aoki Y., Okada T., Yamada H. Formulation of a mathematical model for mechanical bone remodeling process. *JSME International Journal Series C—Mechanical Systems Machine Elements and Manufacturing* 43 (4): 830–836 (2000).
23. Lemaire V., Tobin F. L., Greller L. D., Cho C. R., Suva L. J. Modeling the interactions between osteoblast and osteoclast activities in bone remodeling. *Journal of Theoretical Biology* 229 (3): 293–309 (2004).
24. Wang Y. N., Qin Q. H., Kalyanasundaram S. A theoretical model for simulating effect of parathyroid hormone on bone metabolism at cellular level. *Molecular & Cellular Biomechanics* 6 (2): 101–112 (2009).
25. Frost H. M. Dynamics of bone remodeling. In *Bone biodynamics*, ed. Frost H. M., p. 316. Boston: Little and Brown (1964).
26. Suda T., Takahashi N., Udagawa N., Jimi E., Gillespie M. T., Martin T. J. Modulation of osteoclast differentiation and function by the new members of the tumor necrosis factor receptor and ligand families. *Endocrinology Review* 20 (3): 345–357 (1999).
27. Khosla S. Minireview: The OPG/RANKL/RANK system. *Endocrinology* 142 (12): 5050–5055 (2001).
28. Zaidi M. Skeletal remodeling in health and disease. *Nature Medicine* 13 (7): 791–801 (2007).
29. Fuller K., Lean J. M., Bayley K. E., Chambers T. J. A role for TGF beta(1) in osteoclast differentiation and survival. *Journal of Cell Science* 113 (13): 2445–2453 (2000).
30. Aubin J. E. Advances in the osteoblast lineage. *Biochemistry and Cell Biology—Biochimie et Biologie Cellulaire* 76 (6): 899–910 (1998).
31. Hofbauer L. C., Khosla S., Dunstan C. R., Lacey D. L., Boyle W. J., Riggs B. L. The roles of osteoprotegerin and osteoprotegerin ligand in the paracrine regulation of bone resorption. *Journal of Bone Mineral Research* 15 (1): 2–12 (2000).
32. Heldin C. H., Miyazono K., ten Dijke P. TGF-beta signaling from cell membrane to nucleus through SMAD proteins. *Nature* 390 (6659): 465–471 (1997).
33. McCarthy T. L., Centrella M. Novel links among Wnt and TGF-beta signaling and Runx2. *Molecular Endocrinology* 24 (3): 587–597 (2010).
34. Monroe D. G., McGee-Lawrence M. E., Oursler M. J., Westendorf J. J. Update on Wnt signaling in bone cell biology and bone disease. *Gene* 492 (1): 1–18 (2012).
35. Holmen S. L., Robertson S. A., Zylstra C. R., Williams B. O. Wnt-independent activation of beta-catenin mediated by a Dkk1-Fz5 fusion protein. *Biochemical and Biophysical Research Communications* 328 (2): 533–539 (2005).

36. Qiang Y. W., Chen Y., Brown N., Hu B., Epstein J., Barlogie B., Shaughnessy J. D., Jr. Characterization of Wnt/beta-catenin signalling in osteoclasts in multiple myeloma. *British Journal of Haematology* 148 (5): 726–738 (2010).
37. Parfitt A. M., Chir B. Cellular basis of bone turnover and bone loss—Rebuttal of osteocytic resorption-bone flow theory. *Clinical Orthopaedics and Related Research* 127: 236–247 (1977).
38. Doty S. Morphological evidence of gap junctions between bone cells. *Calcified Tissue International* 33 (1): 509–512 (1981).
39. Shapiro F. Variable conformation of GAP junctions linking bone cells: A transmission electron microscopic study of linear, stacked linear, curvilinear, oval, and annular junctions. *Calcified Tissue International* 61 (4): 285–293 (1997).
40. Nijweide P. J., Burger E. H., Klein-Nulend J. The osteocyte. In *Principles of bone biology*, ed. Bilezikian J. P., Raisz L. G., Rodan G. A., pp. 93–107. San Diego CA: Academic Press (2002).
41. Knothe Tate M. L., Steck R., Forwood M. R., Niederer P. In vivo demonstration of load-induced fluid flow in the rat tibia and its potential implications for processes associated with functional adaptation. *Journal of Experimental Biology* 203 (18): 2737–2745 (2000).
42. Weinbaum S., Cowin S. C., Zeng Y. A model for the excitation of osteocytes by mechanical loading-induced bone fluid shear stresses. *Journal of Biomechanics* 27 (3): 339–360 (1994).
43. Klein-Nulend J., Bacabac R. G., Mullender M. G. Mechanobiology of bone tissue. *Pathologie–Biologie* 53 (10): 576–580 (2005).
44. Ajubi N. E., KleinNulend J., Nijweide P. J., Vrijheid Lammers T., Alblas M. J., Burger E. H. Pulsating fluid flow increases prostaglandin production by cultured chicken osteocytes—A cytoskeleton-dependent process. *Biochemical and Biophysical Research Communications* 225 (1): 62–68 (1996).
45. Mullender M., El Haj A., Yang Y., van Duin M., Burger E., Klein-Nulend J. Mechanotransduction of bone cells in vitro: Mechanobiology of bone tissue. *Medical and Biological Engineering and Computing* 42 (1): 14–21 (2004).
46. Bacabac R. G., Smit T. H., Mullender M. G., Dijcks S. J., Loon J. J. V., Klein-Nulend J. Nitric oxide production by bone cells is fluid shear stress rate dependent. *Biochemical and Biophysical Research Communications* 315 (4): 823–829 (2004).
47. Burger E. H., Klein-Nulend J. Mechanotransduction in bone—Role of the lacuno-canalicular network. *FASEB Journal* 13 (9001): 101–112 (1999).
48. Fan X., Roy E., Zhu L., Murphy T. C., Ackert-Bicknell C., Hart C. M., Rosen C., Nanes M. S., Rubin J. Nitric oxide regulates receptor activator of nuclear factor-{kappa}B ligand and osteoprotegerin expression in bone marrow stromal cells. *Endocrinology* 145 (2): 751–759 (2004).
49. Machwate M., Harada S., Leu C. T., Seedor G., Labelle M., Gallant M., Hutchins S. A., et al. Prostaglandin receptor EP4 mediates the bone anabolic effects of PGE2. *Molecular Pharmacology* 60 (1): 36–41 (2001).
50. Hill A. V. The possible effects of the aggregation of the molecules of hemoglobin on its dissociation curves. *Journal of Physiology* (London) 40 (iv–vii) (1910).
51. Alon U. *An introduction to systems biology: Design principles of biological circuits.* Boca Raton, FL: CRC Press (2007).
52. Roodman G. D. Cell biology of the osteoclast. *Experimental Hematology* 27 (8): 1229–1241 (1999).

53. Hauschka P. Growth factors effects. In *Bone,* vol. 1, ed. Hall, B. K., 512 pp. Boca Raton, FL: CRC Press (1989).
54. Bonewald L. F., Dallas S. L. Role of active and latent transforming growth-factor-beta in bone formation. *Journal of Cellular Biochemistry* 55 (3): 350–357 (1994).
55. Alliston T., Choy L., Ducy P., Karsenty G., Derynck R. TGF-beta-induced repression of CBFA1 by SMAD3 decreases cbfa1 and osteocalcin expression and inhibits osteoblast differentiation. *Embo Journal* 20 (9): 2254–2272 (2001).
56. Schriefer J. L., Warden S. J., Saxon L. K., Robling A. G., Turner C. H. Cellular accommodation and the response of bone to mechanical loading. *Journal of Biomechanics* 38 (9): 1838–1845 (2005).
57. Robling A. G., Burr D. B., Turner C. H. Recovery periods restore mechanosensitivity to dynamically loaded bone. *Journal of Experimental Biology* 204 (19): 3389–3399 (2001).
58. Saxon L. K., Robling A. G., Alam I., Turner C. H. Mechanosensitivity of the rat skeleton decreases after a long period of loading, but is improved with time off. *Bone* 36 (3): 454–464 (2005).
59. Bergmann G., Graichen F., Rohlmann A. Hip joint force measurements. Available from <http://www.medizin.fu-berlin.de/missing.html> (2003).
60. Turner C. H., Forwood M. R., Otter M. W. Mechanotransduction in bone: Do bone cells act as sensors of fluid flow? *FASEB Journal* 8 (11): 875–878 (1994).
61. Warden S. J., Turner C. H. Mechanotransduction in the cortical bone is most efficient at loading frequencies of 5–10 Hz. *Bone* 34 (2): 261–270 (2004).
62. Rubin C., Xu G., Judex S. The anabolic activity of bone tissue, suppressed by disuse, is normalized by brief exposure to extremely low-magnitude mechanical stimuli. *FASEB Journal* 15(12): 2225–2229 (2001).
63. Burr D. B., Robling A. G., Turner C. H. Effects of biomechanical stress on bones in animals. *Bone* 30 (5): 781–786 (2002).
64. Rubin C., Turner A. S., Bain S., Mallinckrodt C., McLeod K. Anabolism. Low mechanical signals strengthen long bones. *Nature* 412 (6847): 603–604 (2001).
65. Klein-Nulend J., van der Plas A., Semeins C. M., Ajubi N. E., Frangos J. A., Nijweide P. J., Burger E. H. Sensitivity of osteocytes to biomechanical stress in vitro. *FASEB Journal* 9 (5): 441–445 (1995).
66. Bakker A. D., Soejima K., Klein-Nulend J., Burger E. H. The production of nitric oxide and prostaglandin E2 by primary bone cells is shear stress dependent. *Journal of Biomechanics* 34 (5): 671–677 (2001).
67. Riggs C. M., Lanyon L. E., Boyde A. Functional associations between collagen fiber orientation and locomotor strain direction in cortical bone of the equine radius. *Anatomy and Embryology* 187 (3): 231–238 (1993).
68. Ascenzi A., Bonucci E. Tensile properties of single osteons. *Anatomical Record* 158 (4): 375–386 (1967).
69. Ascenzi A, Bonucci E. The compressive properties of single osteons. *Anatomical Record* 161 (3): 377–391 (1968).
70. Burr D. B., Hirano T., Turner C. H., Hotchkiss C., Brommage R., Hock J. M. Intermittently administered human parathyroid hormone(1–34) treatment increases intracortical bone turnover and porosity without reducing bone strength in the humerus of ovariectomized cynomolgus monkeys. *Journal of Bone Mineral Research* 16 (1): 157–165 (2001).
71. Wang Y. N., Qin Q. H. Parametric study of control mechanism of cortical bone remodeling under mechanical stimulus. *Acta Mechanica Sinica* 26 (1): 37–44 (2010).

8

Bone Remodeling under Pulsed Electromagnetic Fields and Clinical Applications

8.1 Introduction

This chapter deals with the PEMF devices that have been widely used clinically to treat nonunion fracture, accelerate bone fracture recovery, and slow down osteoporosis. It is an extension of Chapters 6 and 7 to the case of bone remodeling under PEMF. The theoretical and numerical results presented in Wang and Qin [1] are described. Typically, a computational method of system biology is used for analyzing bone remodeling under PEMF at the cellular level, based on experimental findings and recent mathematical advances.

The use of electrical stimulation in bone can be dated back almost 200 years to when a patient with tibia nonunion was successfully cured in 1812 [2]. In 1957, Fukuda and Yasuda [3] discovered the piezoelectric feature of bone, in that when bone was under compression an electronegative potential was induced, whereas an electropositive potential was produced by bone under tension. In 1991, Grande et al. [4] found that the controlling signal seemed to be the electric potential generated by shear piezoelectricity in collagen fibers and/or the streaming potential in canaliculae. These two discoveries raised the possibility that the behavior of bone cells could be affected by externally applied electrical stimuli [5]. Bassett [6] appears to have been the first to use a pair of Helmholtz coils to produce a magnetic field across a fracture site and enhance osteogenesis. Qin and Ye [7], Qin, Qu, and Ye [8], Qin [9], Qu, Qin, and Kang [10], and Qu and Qin [11] studied multifield bone remodeling processes extensively, using the concept of adaptive piezoelectric theory.

Early in the 1990s, several major forms of electrical stimulation were reported to produce osteogenesis, including capacitively and inductively coupled electromagnetic and direct current fields [12,13]. Since then, research into electrically induced osteogenesis in bone has been carried out using these methods both in vivo and in vitro [12,14–16]. The osteogenetic effect on a bone can be used to treat not only long bone fractures [6] but also osteoarthritic joints [17] and osteoporotic bones [18,19], as well as to reverse femoral head necrosis and augment spinal fusion [20].

The biological process involved in the osteogenesis of a bone engendered by PEMF devices is known as cellular bone remodeling. At the cellular level, bone remodeling is an organized process where osteoclasts remove the old bone and osteoblasts replace it with newly formed bone. The osteoclasts and osteoblasts work in a coupled manner within a BMU, which is a mediator mechanism bridging individual cellular activity to whole bone morphology [21] that follows an activation–resorption–formation sequence [22]. As explained in Section 6.2, the RANK–RANKL–OPG pathway [23] provides a clearer picture regarding the control of osteoclastogenesis and bone remodeling in general. The main switch for osteoclastic bone resorption is the RANKL [24], a cytokine that is released by preosteoblasts [25].

Bone cell differentiation and proliferation are important factors during bone remodeling, and clinical PEMF devices have been shown to affect differentiation and proliferation of bone cells in vitro [19,26]. Although it has been proposed that gap junctions, which are specialized intercellular junctions, be considered as mediators of the PEMF-related cellular responses [26–29], the underlying mechanism at cellular level that regulates bone remodeling under PEMF remains poorly understood because of the inconsistent or even contradictory results from experiments. For example, cell proliferation, as assayed by cell number and H-thymidine incorporation, has been reported to increase [30], decrease [14], and remain unaffected [31] by PEMF exposure. Similarly, the production of alkaline phosphatization has been reported either to increase [32] or decrease [28] following PEMF exposure.

In order to remove the limitations to generalization with respect to causes and effects of bone remodeling under PEMF, mathematical models can be used to provide a dynamic, quantitative, and systematic description of the relationships among interacting components of the biological system. Mathematical modeling has been recognized as a powerful tool for testing and analyzing various hypotheses in complex systems that are very difficult (such as time- or money consuming) or just impossible to apply in vivo or in vitro. However, relatively few mathematical models have yet been proposed regarding bone remodeling.

As mentioned in Chapter 6, Kroll [33] and Rattanakul et al. [34] each proposed a mathematical model accounting for the differential activity of PTH administration on bone accumulation. Komarova et al. [35] presented a theoretical model of autocrine and paracrine interactions among osteoblasts and osteoclasts. Komarova [36] also developed a mathematical model that describes the actions of PTH at a single site of bone remodeling, where osteoblasts and osteoclasts are regulated by local autocrine and paracrine factors. Potter et al. [37] proposed a mathematical model for PTH receptor (PTH1R) kinetics, focusing on the receptor's response to PTH dosing to discern bone formation responses from bone resorption. Lemaire et al. [38] incorporated detailed biological information and a RANK–RANKL–OPG pathway into the remodeling cycle of a model that included the catabolic effect of PTH on bone, but the anabolic effect of PTH was not described.

Based on the model of Lemaire et al. [38], Wang, Qin, and Kalyanasundaram [39] developed a mathematical model that could simulate the anabolic behavior of bone affected by intermittent administration of PTH; a theoretical model and its parametric study of the control mechanisms of bone remodeling under mechanical stimulus was also developed by Wang and Qin [40] and Qin and Wang [41]. Pivonka et al. [25,42] extended the bone-cell population model based on the model of Lemaire et al. [38] to explore the model structure of cell–cell interactions theoretically [25], and then they investigated the role of the RANK–RANKL–OPG pathway in bone remodeling [42].

Although many in vitro and in vivo studies have been performed, the cellular mechanism by which PEMF affects bone remodeling is still elusive. To clarify the underlying mechanism at cellular level of regulating the effect of PEMF on bone remodeling, based on the cell population dynamics model [25] and the work reported in Wang et al. [39] and Wang and Qin [40], Wang and Qin [1] developed a mathematical model of bone cell population in bone remodeling under PEMF using the computational method of system biology. Wang and Qin [1] used computational system biology to integrate experimental data into a system-level model, enabling the various interactions to be investigated efficiently and methodically. In particular, the validated model generated using computational system biology can be used as a tool to reduce ambiguity as to causes and effects in complex systems such as bone remodeling, making it possible to test various experimental and theoretical hypotheses in silico [43], and subsequently to develop pharmaceutical and clinical interventions for metabolic bone diseases.

8.2 Model Development

8.2.1 Effects of PEMF on Bone Remodeling

The RANK–RANKL–OPG signaling pathway between osteoblasts and osteoclasts, PTH, and the dual action of TGF-β is diagrammed in Figure 6.1. PEMF applies its effects on bone cells partly through this pathway, and this concept is supported by a number of studies. In an in vitro study [15], a PEMF with a frequency of 15 Hz (1 G [0.1 mT]; electric field strength 2 mV/cm) was applied to neonatal mouse calvarial bone cell cultures for 14 days. The results demonstrated that PEMF stimulation significantly increased the proliferation of osteoblasts. The OPG expression was upregulated and the RANKL concentration was downregulated, compared to the control group.

In another study [44], researchers investigated the effect of PEMF with parameters modified by clinical bone stimulator devices and concluded that OPG might be a potential intermediary involved in the interplay between PEMF stimulation and osteoclastogenesis. Where appropriate, PEMF intensities could either promote or suppress OPG expression in the osteoblastic lineage. Moreover, the osteogenesis effect of PEMF was accompanied by a decrease of RANKL. The research of Schwartz et al. [45] also demonstrated that PEMF induces cells in the osteoblast lineage to express OPG.

Several further studies have demonstrated that PEMF causes osteoblasts to produce other paracrine factors, including TGF-β1, prostaglandin E2 (PGE$_2$), and bone morphogenetic protein-2 [46–48]. Moreover, macrophage colony-stimulating factor (M-CSF) has been shown to decrease after PEMF exposure [44] and bone morphogenetic protein-2, -4, and -5 was found to increase in osteoblasts after PEMF application [49,50]. However, these observations were not consistent in the literature; as a result, they are not included in the effects of PEMF on bone remodeling in the model in this study.

Unlike drug administration, PEMF stimulation can produce a local concentration of growth factor synthesis, without any systemic side effects. It is important, however, to bear in mind that, as with a drug, the dosage of physical stimulus is fundamental if positive effects on osteogenesis are to be produced. The biological effects of PEMF stimulation depend not only on the treatment time but also on signal characteristics such as intensity, waveform, frequency, and length of the signal [51]. The work reported in Zhang et al. [52] indicates that PEMF is the most responsive, compared with other waveforms such as rectangular electromagnetic fields (REMFs), triangular electromagnetic fields (TEMFs), and sinusoidal electromagnetic fields (SEMFs), in terms of their effects on the proliferation and differentiation of osteoblastic cells.

With regard to the different types of PEMF, Bassett, Mitchell, and Gaston [53] proposed that single-pulse is better than burst pulsed PEMF stimulation for osteoporosis prevention and nonunion fracture healing, whereas burst pulsed PEMF stimulation has better effects on acceleration of bone fracture healing. Hannay, Leavesley, and Pearcy [54] examined the response of osteoblast-like cells to a PEMF stimulus, mimicking that of a clinically available device, using four protocols of the timing of the stimulus, each conducted over 3 days. Protocol 1 stimulated the cells for 8 hours each day, protocol 2 stimulated the cells for 24 hours on the first day, protocol 3 stimulated the cells for 24 hours on the second day, and protocol 4 stimulated the cells for 24 hours on the third day. In terms of proliferation and differentiation of the cells compared with the control group, no clear trend was observed between the four protocols.

Intensity of the PEMF is also an important factor, as data from Chang et al. [44] demonstrated that PEMF with different intensities could regulate osteoclastogenesis, bone resorption, OPG, RANKL, and M-CSF concentrations in marrow culture. Specifically, in this experiment, the authors used three different electric field intensities of PEMF (4.8, 8.7, and 12.2 μV/cm) and observed that the recruitment of osteoclast-like cells was inhibited by approximately 33% and increased by approximately 10% when electric field intensities of PEMF were 4.8 and 12.2 μV/cm, respectively. No significant differences across all time points were observed compared with the control group.

8.2.2 Mathematical Model

In the cell population dynamics model presented in Wang and Qin [1], three cell populations (see osteoblastic and osteoclastic lineages in Figure 7.1) are considered in model equations including OBP, OBA, and OCA. OBUs and osteoclastic precursors (OCPs) work as reservoirs whereby the cells differentiate into functional cells such as osteoblasts and osteoclasts, respectively. Their numbers are much larger than the functional cells OBP, OBA, or OCA. As a result, OBU and OCP are assigned a very large constant compared with other cell numbers in the model (i.e., $OBU = OCP = 1 \times 10^{-2}$ pM).

In a manner similar to the treatment in the previous chapter, the Hill equations (7.1), (7.3), and (7.4) are again used to describe the activation and repression of the receptor–ligand interactions.

The equations governing the evolution of the number of osteoblastic and osteoclastic cells in each maturation stage are simply balance equations [38], which means that each cell stage is fed by an entering flow and is emptied by the outgoing flow of differentiated or apoptotic cells (see Figure 8.1). As a result, utilizing Figure 8.1 and based on the formulation in Pivonka et al. [25], the bone cell population dynamics can be formulated as follows:

$$\frac{dOBP}{dt} = D_{OBU} \cdot OBU \cdot \Pi^{T\beta}_{act,OBU} - D_{OBP} \cdot OBP \cdot \Pi^{T\beta}_{rep,OBP} \tag{8.1}$$

$$\frac{dOBA}{dt} = D_{OBP} \cdot OBP \cdot \Pi^{T\beta}_{rep,OBP} - A_{OBA} \cdot OBA \tag{8.2}$$

$$\frac{dOCA}{dt} = D_{OCP} \cdot OCP \cdot \Pi^{RL}_{act,OCP} - A_{OCA} \cdot OCA \cdot \Pi^{T\beta}_{act,OCA} \tag{8.3}$$

$$\frac{dBV}{dt} = k_{for} \cdot \left[OBA(t) - OBA(0) \right] - k_{res} \cdot \left[OCA(t) - OCA(0) \right] \tag{8.4}$$

where D_{OBU}, D_{OBP}, D_{OCP}, A_{OBA}, and A_{OCA}, are explained after Equation (7.9). BV represents bone volume in percentage (%), and k_{for} and k_{res} are the relative bone formation and bone resorption rates, respectively. The simulation starts from a so-called "steady state," whereby BV is 100% and $dBV/dt = 0$, correspondingly; $OBA(t)$ is $OBA(0)$ and $OCA(t)$ is $OCA(0)$. Equations (8.1)–(8.3) are

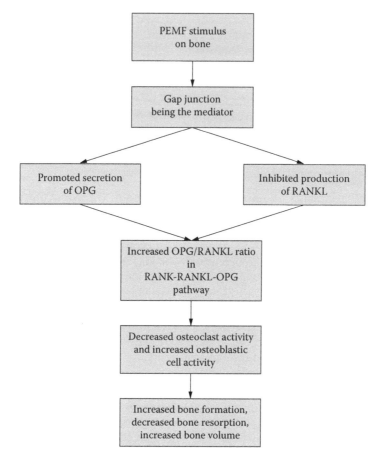

FIGURE 8.1
Schematic diagram of the mathematical model structure of PEMF stimulated bone remodeling at the cellular level.

then transformed as Equations (7.43), (7.44), and (7.46), and the initial values of the model variables have already been listed in Table 7.2.

The model includes two different time scales: a short time scale is used to describe the receptor–ligand reactions such as RANK–RANKL, OPG–RANKL, and TGF-β with its receptor; and a long time scale is required to capture the cell number changes such as OBP, OBA, and OCA. Note that the receptor–ligand reaction is much faster than the changes in cell numbers; therefore, a quasisteady-state assumption is used in the model to describe the receptor–ligand reactions.

Using the short time scale and a quasisteady-state assumption, the expression of TGF-β is obtained and found to be the same in Equation (7.11). Consequently, the activation and repression forms of TGF-β can be obtained and defined by Equations (7.12)–(7.14). The corresponding PTH concentration and its relevant activation and repression functions are also defined by Equations (7.25)–(7.27).

As stated previously, it has already been concluded that PEMF stimulates the production of OPG expressed in OBA and inhibits the expression of RANKL in OBP, based on experimental observations (see Section 8.2.1). OPG and RANKL are the only two factors that are considered to be regulated by PEMF stimulation in the model. The systemic hormone PTH downregulates OPG production in OBA. Therefore, based on Pivonka et al. [25], the OPG concentration can be expressed as

$$OPG = \frac{\beta_{OPG} \cdot OBA \cdot \left(\Pi_{rep,OBA}^{PTH} + F_{OPG}\left(I, f, t, w \right) \right) + P_{OPG,d}\left(t \right)}{\dfrac{\beta_{OPG} \cdot OBA \cdot \left(\Pi_{rep,OBA}^{PTH} + F_{OPG}\left(I, f, t, w \right) \right)}{OPG_{\max}} + \tilde{D}_{OPG}} \tag{8.5}$$

where $F_{OPG}(I, f, t, w)$ is the influence of PEMF on OPG secretion characterized by its intensity, frequency, time, and waveform. Further, PEMF inhibits RANKL expression in OBP and PTH upregulates the RANKL "effective carrying capacity" of OBP [39]. Using the formulation presented in Pivonka et al., [25], the concentration of RANKL can be obtained:

$$RL = \left(\frac{R_{RL} \cdot OBP \cdot \Pi_{act,OBP}^{PTH}}{1 + K_{A1,RL} \cdot OPG + K_{A2,RL} \cdot RK} \right)$$
$$\cdot \left(\frac{\beta_{RL} \cdot OBP \cdot F_{RL}\left(I, f, t, w \right) + P_{RL,d}\left(t \right)}{\beta_{RL} \cdot OBP \cdot F_{RL}\left(I, f, t, w \right) + \tilde{D}_{RL} \cdot R_{RL} \cdot OBP \cdot \Pi_{act,OBP}^{PTH}} \right) \tag{8.6}$$

where $F_{FR}(I, f, t, w)$ is the effect of PEMF on RANKL production characterized by its intensity, frequency, time, and waveform. Then, the activation function of RANKL on differentiation of osteoclast precursor cells OCP can be obtained using Equations (7.3) and (8.6) and is the same as Equation (7.30), except that RL is calculated from Equation (8.6). All the model parameters and their values can be found in Tables 6.1 and 7.1.

8.3 Numerical Investigation of the Model

Bone remodeling is an important biological system when it comes to bone fracture healing, nonunion fracture, and bone diseases such as osteoporosis. It is executed by coordinated activities of osteoclastic cells and osteoblastic cells in BMUs. The coupling between osteoclastic cells and osteoblastic cells is facilitated by the RANK–RANKL–OPG pathway, together with the systemic hormone PTH and transforming growth factor TGF-β. PEMF devices are used clinically to promote bone healing, especially in nonunion fracture, but relatively little is known about the mechanisms involved.

Wang and Qin [1] proposed a mathematical model to simulate the effect of PEMF on bone remodeling at the cellular level, which would help to better understand the underlying mechanisms. In their analysis, Equations (8.1)–(8.4) were numerically solved using Matlab and a series of graphs were plotted for the concentration dynamics of OPG; RANKL; cell populations of OBA, OCA, and OBP; and bone volume, as shown in Figures 8.2–8.5, respectively. Note that the parameter values listed in Tables 6.1 and 7.1 have been used in the analysis.

The effects of PEMF on bone remodeling are dependent on its intensity, frequency, waveform, and application time. According to Hannay et al. [54], the duration of the PEMF stimulation does not affect bone cell development, which is different from bone remodeling under mechanical stimulus. In this numerical investigation, the specific parameter values of PEMF were chosen from the widely used PEMF devices in clinics [55]; this set of parameter values is the only one used in this study's model. Consequently, it is assumed that the effects of PEMF on bone remodeling in this model, specifically OPG or RANKL, do not change (represented by two different constants, F_{OPG} and F_{RL}, in the model) throughout the simulation (3 months).

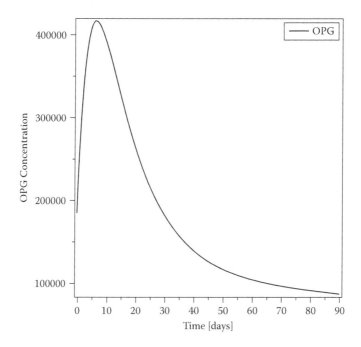

FIGURE 8.2
OPG concentration dynamics during 3-month PEMF application.

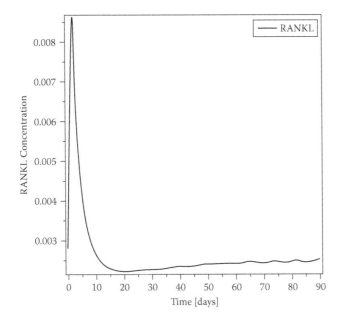

FIGURE 8.3
RANKL concentration dynamics during 3-month PEMF application.

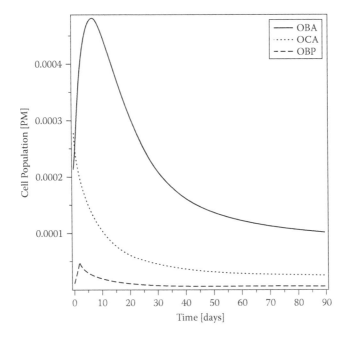

FIGURE 8.4
OBA, OCA, and OBP cell population dynamics during 3-month PEMF application.

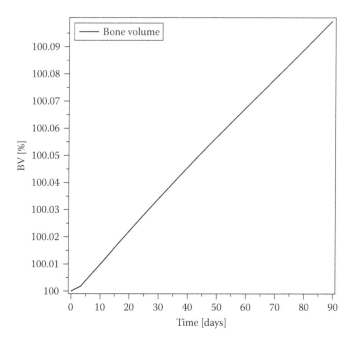

FIGURE 8.5
Bone volume percentage dynamics during 3-month PEMF application.

The concentration dynamics of OPG during the simulated 3-month PEMF application are shown in Figure 8.2. Consistent with experimental observations, the OPG concentration increases during the first 10 days' simulation. Surprisingly, in the simulation, the concentration of OPG starts to drop from the 10th day. Because most in vitro experiments have been performed over 2 weeks, there is no available experimental data for comparison. The pattern of OPG concentration might be explained by the fact that because of the stimulus effect of the PEMF, OPG concentration increases for a relatively short period after the onset of PEMF application, and then the binding of OPG with RANKL catches up as bone remodeling occurs. More OPGs are consumed than produced by osteoblast cells; as a result, OPG concentration drops until the end of the simulation.

Figure 8.3 shows the RANKL concentration dynamics during the 3-month PEMF application. As can be observed from the graph, within 2 weeks the RANKL concentration drops, compared with the initial value. However, it was not expected that the RANKL concentration would significantly increase immediately after the simulation began, followed by a dramatic decrease and the maintenance of a similar level throughout the rest of the simulation. A possible reason for this development pattern is that the number of preosteoblast cells that produce RANKL is increased by the stimulus effect of PEMF on the proliferation of bone marrow mesenchymal stem cells [16] immediately after the application of PEMF. Then, the inhibitory effect of PEMF on the OBP takes over and this overall trend is maintained until the end of the simulation.

The cell population dynamics of OBA, OCA, and OBP can be observed in Figure 8.4. As expected, the OBA and OBP populations rise and OCA decreases in the first 2 weeks of the simulation, which is consistent with experimental observations. Due to the coupling effect between OCA and OBA, the OBA population starts to decrease after it peaks and continues to decrease until the end of the simulation, while maintaining a higher concentration than OCA, which accounts for the continuing growth of bone volume in Figure 8.5.

8.4 Parametric Study of Control Mechanism of Bone Remodeling under PEMF

Based on the mathematical model presented before, an extensive parametric study was performed [1]. Model parameters related to fundamental cell behaviors such as differentiation and apoptosis were investigated, in order to identify putative control mechanisms for physiologically reasonable bone remodeling under PEMF.

The functional outputs of the bone remodeling system, such as bone loss or gain, or homeostasis, are executed by BMUs whereby osteoclasts absorb bone mineral in bone matrix and activated osteoblasts lay down the newly formed bone. The BMU acts as a mediator mechanism, bridging individual cellular activity to whole bone morphology [21], and that mechanism is sensitive to any changes in its microenvironment. It is expected, therefore, that modification to any component of the BMU will have a significant effect on its output behavior.

From a control theory perspective, one can always argue that there must be several control mechanisms working simultaneously in the complex bone remodeling system under PEMF, governing the response of a BMU to changes in its microenvironment by modifying the differentiation or apoptosis rates of bone cells. Wang and Qin [1] applied perturbations to the bone remodeling system under PEMF (which is in a steady state) by down- and upregulating its parameters in random combination groups of five differentiation and apoptosis rate parameters: DF_{obu}, DF_{ocp}, DF_{obp}, A_{oba}, and A_{oca}. Each parameter in each group (groups of one, two, three, four, and finally all five parameters at one time) could be up- or downregulated. Using simple combination theory, the total number of permutation could be calculated and is

$$242 = \sum_{i=1}^{5} C_5^i \cdot 2^i$$

Then, in order to investigate the system behavior for a wide range of changes, the exponentially changed factor (1.5^{ex}) was applied to each of the five differentiation and apoptosis rate parameters, whereby the exponent *ex* ranged from –10 to 10 with a step increase of 0.5. The assessment of the effect of each of the parameter combinations to the system behavior was chosen as the responses of bone volume, which were sampled on the 90th day to indicate the maximum change. Using Matlab, all 242 graphs were plotted. Then, summarizing all the plots of bone volume versus variation of exponent *ex*, Wang and Qin [1] found three subsets of curves, which are plotted in Figure 8.6.

Figure 8.6(a) and 8.6(b) demonstrates an exponential increase and decrease of bone volume, respectively, for increasing the model parameter exponentially (exponent e^x from –10 to 10). This type of behavior is considered physiologically unrealistic from a biological perspective and was obtained for quite a large range of model parameter combinations. Conversely, Figure 8.6(c)

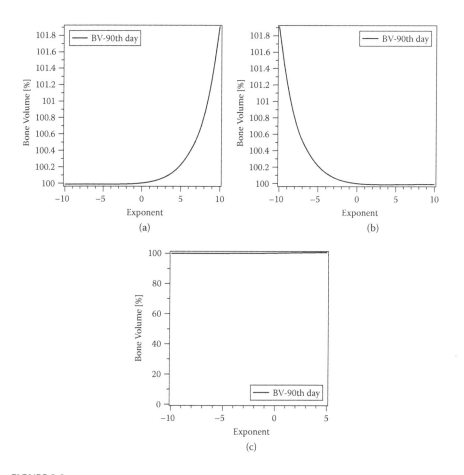

FIGURE 8.6
Physiologically unrealistic changes of BMC (bone mineral content) and BFE (bone fracture energy) versus combined changes of model parameter $[1.5^{-10} - 1.5^{+10}]$ *p:* (a) exponential bone growth; (b) exponential bone decrease; and (c) slight changes of bone. (*p* is the parameter value.)

represents the other extreme, wherein only minor changes of bone volume occur during the entire range of parameter variation. These three types of response curve were excluded from further analysis because they did not provide an effective control mechanism for bone volume.

In Pivonka et al. [25], the "idealized" regulatory response by functionally active BMUs is discussed. As mentioned previously, the bone remodeling system starts from a steady state, in which it can be identified that $\Delta BV = 0$, and concentrations of various hormones and growth factors cause initial values of differentiation and apoptosis rates in BMUs. In order to respond to minor changes in concentrations, it is expected that BMUs should be rather insensitive to these fluctuations. Referring to Figure 7.7, point A can be identified as the concentration threshold, which means that any change in a model parameter less than A causes no change in bone volume. Further, a region around the usual operational status of BMUs should be found with relatively small gradients of changes in bone volume in response to changes in differentiation rates (regions C–D and D–E in Figure 7.7), and a region with larger gradients for larger changes in differentiation rates (region E–F in Figure 7.7). It is expected that this response in bone volume change will remain limited if the differentiation rates increase significantly (the region beyond point F in Figure 7.7) because an unlimited rise in bone volume is not physiologically realistic.

Conversely, it is expected that the rate of bone volume change will also decrease in a limited manner if the differentiation rates decrease significantly. In fact, it is physiologically reasonable for the bone volume change to be zero for extremely small differentiation rates. It can also be observed from Figure 7.7 that point F marks the maximum change in bone volume (ΔB_{max}). Since point A is the maximum concentration that does not lead to further modification of bone volume, there must be a transition region from point C to A, which is characterized by point B, the lowest value of bone volume.

Having detailed a potential "ideal response curve," the search for response curves that might meet these requirements can now begin. Encouragingly, a small number of curves are identified that bear similarity to the ideal response curve. Table 8.1 summarizes all the parameter combinations that produce idealized response curves. In Figure 8.7, the physiologically realistic response curve is plotted, which corresponds to the parameter permutation involving three parameters (A_{OBA}, A_{OCA}, $A_{OST} = -/+/+$) and is similar to the ideal response curve shown in Figure 7.7.

8.5 Effects of PEMF on Patients Undergoing Hip Revision

In the previous sections of this chapter, theoretical and numerical analyses due to the presence of electromagnetic field were discussed. In this section, clinical applications of electromagnetic fields in bone healing are briefly

TABLE 8.1

Summary of Parameter Combinations That Lead to a Controlled Remodeling Process

Number of Parameters in a Combination	Combinations of Differentiation and Apoptosis Rates	Variation of Each Parameter
2	D_{OBP}/A_{OBA}	+/+
2	D_{OCP}/A_{OST}	+/−
2	A_{OBP}/A_{OCA}	−/+
3	$D_{OBP}/A_{OBA}/A_{OCA}$	−/+/−
3	$D_{OBP}/A_{OBA}/A_{OCA}$	−/−/+
3	$D_{OBP}/A_{OBA}/A_{OST}$	+/−/+
3	$A_{OBA}/A_{OCA}/A_{OST}$	−/+/+
4	$D_{OBP}/A_{OCP}/A_{OBA}/A_{OCA}$	−/−/−/−
4	$D_{OBP}/D_{OCP}/A_{OBA}/A_{OCA}$	−/−/−/+
4	$D_{OBP}/A_{OBA}/A_{OCA}/A_{OST}$	−/+/+/−
5	$D_{OBU}/D_{OBP}/D_{OCP}/A_{OBA}/A_{OST}$	+/−/−/+/−
5	$D_{OBU}/D_{OBP}/A_{OBA}/A_{OCA}/A_{OST}$	−/+/−/+/−
5	$D_{OBP}/D_{OCP}/D_{OBA}/A_{OCA}/A_{OST}$	−/−/−/−/−
5	$D_{OBP}/D_{OCP}/D_{OBA}/A_{OCA}/A_{OST}$	+/−/−/+/−

Note: Variations with "+" represent a parameter increase and those with "−" represent a parameter decrease.

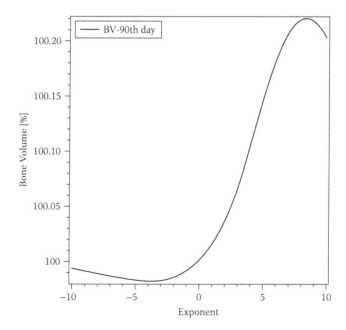

FIGURE 8.7
Typical physiologically realistic fluctuations of bone volume with combinations of parameter change.

described. The use of magnetic field therapy in clinical applications dates back over 500 years. In the fifteenth century, Swiss physician and alchemist Philippus von Hohenheim (known as Paracelsus) used lodestones, or naturally magnetized pieces of the mineral magnetite, to treat conditions such as epilepsy, diarrhea, and hemorrhage. He believed that the ability of magnets to attract iron could be replicated by attracting disease away from the body.

As described in Wikipedia (http://en.wikipedia.org/wiki/Pulsed_electromagnetic_field_therapy; accessed March 2011), pulsed electromagnetic field therapy is a reparative technique most commonly used in the field of orthopedics for the treatment of nonunion fractures, failed fusions, congenital pseudarthrosis, and depression. In the case of bone healing, PEMF uses electrical energy to direct a series of magnetic pulses through injured tissue whereby each magnetic pulse induces a tiny electrical signal that stimulates cellular repair. Many studies have also demonstrated the effectiveness of PEMF in healing soft-tissue wounds, suppressing inflammatory responses at the cell membrane level to alleviate pain, and increasing range of motion. The value of pulsed electromagnetic field therapy has been shown to cover a wide range of conditions, with well-documented trials carried out by hospitals, rheumatologists, physiotherapists, and neurologists. The emphasis in this section, however, is focused on the developments in Dallari et al. [56].

8.5.1 Basic Process

Dallari et al. [56] considered a group of 30 subjects who had undergone hip prosthesis revision. Exclusion criteria were the presence of autoimmune disease, diabetes mellitus, cancer, infectious disease, and lymphoproliferative disorder. Subjects were assigned either to the active or to the placebo group according to a computer-generated schedule: A random number seed was entered into the computer to generate a list that assigned equal numbers of active and placebo stimulators (blocks of four: two active and two placebos). Subjects and the medical staff were unable to differentiate between the active and placebo stimulators. The characteristics of the stimulators were made known only after all clinical and instrumental evaluations had been completed.

As Dallari et al. described, for all subjects the assessment of the mobilization of stems was performed on x-ray films according to the Gustilo-Pasternak classification, which classifies femoral component loosening into four types based on the severity of loosening and instability: type I: minimal endosteal or inner cortical bone loss

  type II: proximal canal enlargement with cortical thinning of 50% or more and sometimes a lateral wall defect with an intact circumferential wall type III: posteromedial wall defect involving the lesser trochanter

type IV: total proximal circumferential bone loss at various distances below the lesser trochanter

The transfemoral approach used involves diaphyseal osteotomy of the femur by an "open book" procedure. This technique allows easy removal of the prosthesis and all the residual cement. After resection, following the direction of the fibers, the gluteus maximus and fascia lata, the greater trochanter, and the vastus lateralis are exposed and then the femoral osteotomy can be performed. The length of the osteotomy corresponds to the primary prosthesis and is delimited by Steinmann nails to evaluate the distance between the osteotomy and the greater trochanter. The osteotomy is performed with a chisel and an oscillating saw which allows the "book opening" of the femur, saving the proximal insertion of the vastus lateralis.

Once the bone is open and the prosthesis exposed, the cement and granulation tissue can be accurately removed until femoral bone bleeding occurs. The shortest stem that ensures sufficient biomechanical stability is implanted in the conically reamed osseous bed until stable stem anchoring is achieved. Subjects were instructed to avoid weight bearing for the first 30 days after the operation; then, partial weight bearing was allowed and, finally, full load bearing was permitted at 90 days after the operation.

8.5.2 Clinical and Densitometric Evaluation

The evaluation conducted by Dallari et al. was as follows [56]. First, subjects were evaluated at the preoperative exam and at the final follow-up (90 days after surgery) using the Merle D'Aubigné scale, modified by Charley. This scoring method includes the evaluation of pain, the ability to walk, and mobility, with each parameter scored on a scale from 1 (pain: severe, also at rest; walking: impossible; mobility: flexion < 15°) to 6 (pain: absent; walking: normal; mobility: flexion > 90°, abduction ≥ 15°).

For densitometric measurement, subjects underwent dual-energy x-ray absorptiometry (DXA) postoperatively, within 10 days, and at 90 days after surgery. DXA was performed using an Eclipse (Norland, Fort Atkinson, Wisconsin) [56]. The subjects were positioned with the limb to be investigated extended and in neutral rotation, using a specific device for limb immobilization. On the DXA image, five regions of interest were identified (see Figure 8.8): two corresponding to the lateral diaphyseal osteotomy from the apex of the greater trochanter, one at the distal osteotomy area, and two at the medial periprosthetic cortex. Region 1 was from the tip of the greater trochanter to a point on the lateral cortex one-third of the distance to the tip of the prosthesis; regions 2 and 3 were immediately distal to region 1 on the lateral aspect of the femur. Region 5 had the same dimension as region 3, but on the medial side, and region 6 was between region 5 and the lower edge of the lesser trochanter.

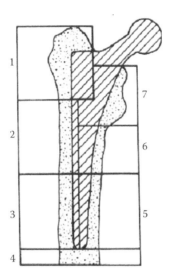

FIGURE 8.8
Schematic representation of a femur implanted with a hip prosthesis [56].

In analyzing the densitometry findings, Dallari et al. subtracted for each area the postoperative DXA values from those measured at 90 days and considered all results above 3.5% as responders; this value was selected on the basis of the coefficient of variation of Eclipse (3%) for this type of investigation.

8.5.3 PEMF Stimulation

Dallari et al. [56] started stimulation within 7 days of the operation and maintained it up to day 90. The subjects were instructed to use the devices for at least 6 hours per day. All the stimulators were equipped with a timer to record the hours of use. The electromagnetic field was generated by a single coil that was positioned on the lateral surface of the thigh and kept in place by a Velcro strap. The pulse generator supplied the coil with single voltage pulses at 75 Hz, each lasting 1.3 ms. The peak amplitude of magnetic field was 2 ± 0.2 mT. This measurement was made by the Hall probe of the gauss-meter [56]. The peak voltage in a coil probe connected to a Tektronix 720A oscilloscope was 2.5 ± 0.2 mV. Since the coil probe consisted of 50 turns and was of 0.5 cm internal diameter, this corresponded to a dB/dt max value of 2.5 T/s.

8.5.4 Discussion

In Dallari et al. [56], PEMF stimulation was applied in humans to enhance the outcome of hip arthroplasty as it can influence a relatively large area and the whole stem can be irradiated homogeneously. PEMFs have been used with success in patients with a loosened prosthesis and with a primary painful uncemented prosthesis [57]. In their randomized prospective double-blind study [56], Dallari et al. focused on the short-term effects of PEMF stimulation after hip prosthesis revision. In fact, if a role has to be identified for PEMF stimulation of the periprosthetic osteogenesis, the effects should be observed in the early days after surgery. PEMF should be used to increase spontaneous osseous regeneration, to limit inflammatory response, to accelerate hip fixation, and to shorten the subject's functional recovery time.

One limitation of the study is that only short-term results are reported. Nevertheless, the initial fixation of the implant is crucial for the success of prosthesis implantation [58]. Also, with strategies based on bisphosphonate treatments that seem to decrease peri-implant bone loss, differences between treated and control subjects have become evident after 3 months [58]. It has also been shown that significant bone loss, up to 14%, occurs during the first 3 months after total hip arthroplasty [58].

Dallari et al. [56] indicated that, even though their double-blind study was performed on a small but homogeneous group of subjects, they were able to show the changes that occurred in bone after only 90 days, as well as the positive effect of PEMF stimulation on BMD, on pain, and, consequently, on functional recovery of subjects. The treatment was not associated with any negative side effects; however, they noted that the use of this electromagnetic stimulation at the hip requires considerable subject commitment. Nevertheless, they suggested that the use of PEMFs should always be considered after revision surgery, especially with severely loosened prostheses and debilitated elderly subjects.

Finally, it is noted that many thousands of people around the world have used PEMF successfully for a wide range of pain-related conditions. In fact, PEMF influences cell behavior by inducing electrical changes around and within the cell wall membrane.

In particular, pulsed electromagnetic wave field therapy creates ionic migration of essential chemicals and proteins from cells into the bloodstream by agitating the cell wall membranes; this subsequently encourages the release of neurochemicals that occur naturally in the brain into the bloodstream. These endorphins have powerful analgesic properties, as do enkaphalins, an endorphin with opiate qualities that occurs in the brain. PEMF also helps to enhance the cell utilization process in the body, which in turn stimulates improved blood circulation, blood oxygen content, cell growth, and blood vessel formation.

In bone fractures, affected tissues generate small electric charges that are greater than those of less stressed matter, so osteoblasts (polarized bone-laying cells) are attracted to these areas and begin to build up extra bone material to counter the stress. PEMF encourages this process.

Further, within the bone itself, the pulsed electromagnetic wave fields delivered induce small eddy currents in the trace elements. In turn, these purify and strengthen the crystal structures that attract bone cells to the area under treatment, thereby accelerating the bone healing process to facilitate earlier mobilization and eventual full bone union.

References

1. Wang Y. N., Qin Q. H. A theoretical study of bone remodeling under PEMF at cellular level. *Computer Methods in Biomechanics and Biomedical Engineering* 15 (8): 885–897 (2012).

2. Brighton C., Magnusson P. Electrically induced osteogenesis. Its clinical use in treating nonunion. In *Bioelectric repair and growth*, ed. Fukuda E., pp. 3–19. Niigata, China: Nisimura (1985).

3. Fukada E., Yasuda I. On the piezoelectric effect of bone. *Journal of the Physical Society of Japan* 12 (10): 1158–1162 (1957).

4. Grande D. A., Magee F. P., Weinstein A. M., McLeod B. R. The effect of low-energy combined AC and DC magnetic-fields on articular-cartilage metabolism. *Annals of the New York Academy of Sciences* 635: 404–407 (1991).

5. Bassett C. A. L., Pawluk R. J., Becker R. O. Effects of electric currents on bone in vivo. *Nature* 204 (495): 652–654 (1964).

6. Bassett C. A. L. Pulsing electromagnetic fields—A new method to modify cell behavior in calcified and non-calcified tissues. *Calcified Tissue International* 34 (1): 1–8 (1982).

7. Qin Q. H., Ye J. Q. Thermoelectroelastic solutions for internal bone remodeling under axial and transverse loads. *International Journal of Solids Structure* 41 (9–10): 2447–2460 (2004).

8. Qin Q. H., Qu C. Y., Ye J. Q. Thermo electroelastic solutions for surface bone remodeling under axial and transverse loads. *Biomaterials* 26 (33): 6798–6810 (2005).

9. Qin Q. H. Multi-field bone remodeling under axial and transverse loads. In *New research on biomaterials*, ed. Boomington D. R., pp. 49–91. New York: Nova Science Publishers (2007).

10. Qu C. Y., Qin Q. H., Kang Y. L. A hypothetical mechanism of bone remodeling and modeling under electromagnetic loads. *Biomaterials* 27 (21): 4050–4057 (2006).

11. Qu C. Y., Qin Q. H. Evolution of bone structure under axial and transverse loads. *Structural Engineering and Mechanics* 24 (1): 19–29 (2006).

12. Kubota K., Yoshimura N., Yokota M., Fitzsimmons R. J., Wikesjo U. M. E. Overview of effects of electrical stimulation on osteogenesis and alveolar bone. *Journal of Periodontology* 66 (1): 2–6 (1995).

13. Mammi G. I., Rocchi R., Cadossi R., Massari L., Traina G. C. The electrical stimulation of tibial osteotomies—Double-blind study. *Clinical Orthopaedics and Related Research* (288): 246–253 (1993).

14. Fredericks D. C., Nepola J. V., Baker J. T., Abbott J., Simon B. Effects of pulsed electromagnetic fields on bone healing in a rabbit tibial osteotomy model. *Journal of Orthopaedic Trauma* 14 (2): 93–100 (2000).
15. Chang W. H., Chen L. T., Sun J. S., Lin F. H. Effect of pulse-burst electromagnetic field stimulation on osteoblast cell activities. *Bioelectromagnetics* 25 (6): 457–465 (2004).
16. Sun L. Y., Hsieh D. K., Yu T. C., Chiu H. T., Lu S. F., Luo G. H., Kuo T. K., Lee O. K., Chiou T. W. Effect of pulsed electromagnetic field on the proliferation and differentiation potential of human bone marrow mesenchymal stem cells. *Bioelectromagnetics* 30 (4): 251–260 (2009).
17. Trock D. H., Bollet A. J., Markoll R. The effect of pulsed electromagnetic fields in the treatment of osteoarthritis of the knee and cervical spine—Report of randomized, double-blind, placebo-controlled trials. *Journal of Rheumatology* 21 (10): 1903–1911 (1994).
18. Chang K., Chang W. H. S. Pulsed electromagnetic fields prevent osteoporosis in an ovariectomized female rat model: A prostaglandin E-2-associated process. *Bioelectromagnetics* 24 (3): 189–198 (2003).
19. Chang K., Chang W. H. S., Wu M. L., Shih C. Effects of different intensities of extremely low frequency pulsed electromagnetic fields on formation of osteoclast-like cells. *Bioelectromagnetics* 24 (6): 431–439 (2003).
20. Linovitz R. J., Pathria M., Bernhardt M., Green D., Law M. D., McGuire R. A., Montesano P. X., et al. Combined magnetic fields accelerate and increase spine fusion—A double-blind, randomized, placebo controlled study. *Spine* 27 (13): 1383–1388 (2002).
21. Frost H. M. *Intermediary organization of the skeleton.* Boca Raton, FL: CRC Press (1986).
22. Robling A. G., Castillo A. B., Turner C. H. Biomechanical and molecular regulation of bone remodeling. *Annual Review of Biomedical Engineering* 8: 455–498 (2006).
23. Anderson D. M., Maraskovsky E., Billingsley W. L., Dougall W. C., Tometsko M. E., Roux E. R., Teepe M. C., et al. A homologue of the TNF receptor and its ligand enhance T-cell growth and dendritic-cell function. *Nature* 390 (6656): 175–179 (1997).
24. Zaidi M. Skeletal remodeling in health and disease. *Nature Medicine* 13 (7): 791–801 (2007).
25. Pivonka P., Zimak J., Smith D. W., Gardiner B. S., Dunstan C. R., Sims N. A., Martin J. T., Mundy G. R. Model structure and control of bone remodeling: A theoretical study. *Bone* 43 (2): 249–263 (2008).
26. Lohmann C. H., Schwartz Z., Liu Y., Li Z., Simon B. J., Sylvia V. L., Dean D. D., et al. Pulsed electromagnetic fields affect phenotype and connexin 43 protein expression in MLO-Y4 osteocyte-like cells and ROS 17/2.8 osteoblast-like cells. *Journal of Orthopaedic Research* 21 (2): 326–334 (2003).
27. McLeod K. J., Rubin C. T. The effect of low-frequency electrical fields on osteogenesis. *Journal of Bone and Joint Surgery, American* vol. 74A (6): 920–929 (1992).
28. Vander Molen M. A., Donahue H. J., Rubin C. T., McLeod K. J. Osteoblastic networks with deficient coupling: Differential effects of magnetic and electric field exposure. *Bone* 27 (2): 227–231 (2000).
29. Tabrah F., Hoffmeier M., Gilbert F., Batkin S., Bassett C. A. L. Bone-density changes in osteoporosis-prone women exposed to pulsed electromagnetic fields (PEMFs). *Journal of Bone Mineral Research* 5 (5): 437–442 (1990).

30. De Mattei M., Caruso A., Traina G. C., Pezzetti F., Baroni T., Sollazzo V. Correlation between pulsed electromagnetic fields exposure time and cell proliferation increase in human osteosarcoma cell lines and human normal osteoblast cells in vitro. *Bioelectromagnetics* 20 (3): 177–182 (1999).

31. Diniz P., Shomura K., Soejima K., Ito G. Effects of pulsed electromagnetic field (PEMF) stimulation on bone tissue like formation are dependent on the maturation stages of the osteoblasts. *Bioelectromagnetics* 23 (5): 398–405 (2002).

32. McLeod K. J., Collazo L. Suppression of a differentiation response in MC-3T3-E1 osteoblast-like cells by sustained, low-level, 30 Hz magnetic-field exposure. *Radiation Research* 153 (5): 706–714 (2000).

33. Kroll M. Parathyroid hormone temporal effects on bone formation and resorption. *Bulletin of Mathematical Biology* 62 (1): 163–188 (2000).

34. Rattanakul C., Lenbury Y., Krishnamara N., Wolwnd D. J. Modeling of bone formation and resorption mediated by parathyroid hormone: response to estrogen/PTH therapy. *Biosystems* 70 (1): 55–72 (2003).

35. Komarova S. V., Smith R. J., Dixon S. J., Sims S. M., Wahl L. M. Mathematical model predicts a critical role for osteoclast autocrine regulation in the control of bone remodeling. *Bone* 33 (2): 206–215 (2003).

36. Komarova S. V. Mathematical model of paracrine interactions between osteoclasts and osteoblasts predicts anabolic action of parathyroid hormone on bone. *Endocrinology* 146 (8): 3589–3595 (2005).

37. Potter L. K., Greller L. D., Cho C. R., Nuttall M. E., Stroup G. B., Suva L. J., Tobin F. L. Response to continuous and pulsatile PTH dosing: A mathematical model for parathyroid hormone receptor kinetics. *Bone* 37 (2): 159–169 (2005).

38. Lemaire V., Tobin F. L., Greller L. D., Cho C. R., Suva L. J. Modeling the interactions between osteoblast and osteoclast activities in bone remodeling. *Journal of Theoretical Biology* 229 (3): 293–309 (2004).

39. Wang Y. N., Qin Q. H., Kalyanasundaram S. A theoretical model for simulating effect of parathyroid hormone on bone metabolism at cellular level. *Molecular & Cellular Biomechanics* 6 (2): 101–112 (2009).

40. Wang Y. N., Qin Q. H. Parametric study of control mechanism of cortical bone remodeling under mechanical stimulus. *Acta Mechanica Sinica* 26 (1): 37–44 (2010).

41. Qin Q. H., Wang Y. N. A mathematical model of cortical bone remodeling at cellular level under mechanical stimulus. *Acta Mechanica Sinica* (submitted) (2012).

42. Pivonka P., Zimak J., Smith D. W., Gardiner B. S., Dunstan C. R., Sims N. A., Martin J. T., Mundy G. R. Theoretical investigation of the role of the RANK–RANKL–OPG system in bone remodeling. *Journal of Theoretical Biology* 262 (2): 306–316 (2010).

43. Defranoux N. A., Stokes C. L., Young D. L., Kahn A. J. In silico modeling and simulation of bone biology: A proposal. *Journal of Bone Mineral Resource* 20 (7): 1079–1084 (2005).

44. Chang K., Chang W. H-S., Huang S., Huang S., Shih C. Pulsed electromagnetic fields stimulation affects osteoclast formation by modulation of osteoprotegerin, RANK ligand and macrophage colony-stimulating facto. *Journal of Orthopaedic Research* 23 (6): 1308–1314 (2005).

45. Schwartz Z., Fisher M., Lohmann C., Simon B., Boyan B. Osteoprotegerin (OPG) Production by cells in the osteoblast lineage is regulated by pulsed electromagnetic fields in cultures grown on calcium phosphate substrates. *Annals of Biomedical Engineering* 37 (3): 437–444 (2009).

46. Bodamyali T., Bhatt B., Hughes F. J., Winrow V. R., Kanczler J. M., Simon B., Abbott J., Blake D. R., Stevens C. R. Pulsed electromagnetic fields simultaneously induce osteogenesis and upregulate transcription of bone morphogenetic proteins 2 and 4 in rat osteoblastsin in vitro. *Biochemical and Biophysical Research Communications* 250 (2): 458–461 (1998).

47. Guerkov H. H., Lohmann C. H., Liu Y., Dean D. D., Simon B. J., Heckman J. D., Schwartz Z., Boyan B. D. Pulsed electromagnetic fields increase growth factor release by nonunion cells. *Clinical Orthopaedics and Related Research* 384:265–279 (2001).

48. Lohmann C. H., Schwartz Z., Liu Y., Guerkov H., Dean D. D., Simon B., Boyan B. D. Pulsed electromagnetic field stimulation of MG63 osteoblast-like cells affects differentiation and local factor production. *Journal of Orthopaedic Research* 18 (4): 637–646 (2000).

49. Nagai M., Ota M. Pulsating electromagnetic-field stimulates messenger—RNA expression of bone morphogenetic protein-2 and protein-4. *Journal of Dental Research* 73 (10): 1601–1605 (1994).

50. Yajima A., Ochi M., Hirose Y., Nakade O., Abiko Y., Kaku T., Sakaguchi K. Effects of pulsing electromagnetic fields on gene expression of bone morphogenetic proteins in human osteoblastic cell line in vitro. *Journal of Bone Mineral Research* 11: T326–T326 (1996).

51. Massari L., Caruso G., Sollazzo V., Setti S. Pulsed electromagnetic fields and low intensity pulsed ultrasound in bone tissue. *Clinical Cases in Mineral and Bone Metabolism* 6 (2): 149–154 (2009).

52. Zhang X., Zhang J., Qu X., Wen J. Effects of different extremely low-frequency electromagnetic fields on osteoblasts. *Electromagnetic Biology and Medicine* 26 (3): 167–177 (2007).

53. Bassett C. A., Mitchell S. N., Gaston S. R. Treatment of ununited tibial diaphyseal fractures with pulsing electromagnetic fields. *Journal of Bone and Joint Surgery* 63 (4): 511–523 (1981).

54. Hannay G. G., Leavesley D. I., Pearcy M. J. Timing of pulsed electromagnetic field stimulation does not affect the promotion of bone cell development. *Bioelectromagnetics* 26 (8): 670–676 (2005).

55. Li J. K. J., Lin J. C. A., Liu H. C., Chang W. H. S. Cytokine release from osteoblasts in response to different intensities of pulsed electromagnetic field stimulation. *Electromagnetic Biology and Medicine* 26 (3): 153–165 (2007).

56. Dallari D., Fini M., Giavaresi G., Del Piccolo N., Stagni C., Amendola L., Rani N., Gnudi S., Giardino R. Effects of pulsed electromagnetic stimulation on patients undergoing hip revision prostheses: A randomized prospective double-blind study. *Bioelectromagnetics* 30 (6): 423–430 (2009).

57. Konrad K., Sevcic K., Foldes K., Piroska E., Molnar E. Therapy with pulsed electromagnetic fields in aseptic loosening of total hip protheses: A prospective study. *Clinical Rheumatology* 15: 325–328 (1996).

58. Peter B., Gauthier O., Laib S., Bujoli B., Guicheux J., Janvier P., van Lenthe G. H., et al. Local delivery of bisphosphonate from coated orthopedic implants increases implants mechanical stability in osteoporotic rats, *Journal of Biomedical Materials Research Part A* 76A: 133–143, 2006.

9

Experiments

9.1 Introduction

In the previous chapters, theoretical and numerical models for bone remodeling were described. Experimental approaches are also important in bone performance analysis. In this chapter some recent developments in experiments with bone materials are reported.

Many experimental studies have been carried out over the past decades on the mechanical behavior of bone tissues under various external loadings [1–4]. These studies have focused on sample preparation, effects of microstructure on bone properties, and piezoelectric properties of bone materials. Yin et al. [5] indicated that the ability to remove soft tissue from skeletons effectively without compromising surface morphology or bone integrity is paramount. Their experiments showed that there was no significant difference among the microstructures in terms of porosity and microhardness values of bones when they were cleaned in water, trypsin, and detergent macerations. In general, the commonly used methods for cleaning bones can be categorized as manual cleaning, enzymatic maceration, cooking, water maceration, and insect consumption [6]. Enzymatic maceration is considered the most convenient method, employing digestive enzymes such as trypsin, prepsin, or papain [7,8]. To enhance cleaning capability, cheap and easily available laundry detergents are also introduced at room temperature or elevated temperatures [6]. Studies have shown that cleaning using powdered detergent allowed the largest segments of DNA to be amplified. It may have less degradative effect on bone DNA than boiling bone and bleaching processes [9,10].

Regarding the biomechanical testing of bone tissue, work to date has mainly focused on macroscopic fracture based on structural characterization [11] and micromechanical behaviors [3]. Studies [12,13] have quantitatively addressed fracture behaviors associated with the microstructure of bone tissues. Relationships between fatigue or cyclic loading and microcrack formation have been reported [11,14]. On the other hand, indentation tests represent a promising direction in characterizing the quality of bone repair as a function of location [11,15–17]. These tests have been used to measure the local mechanical properties of callus and surrounding tissues [17–19]. Multifield measurement of bone tissues has also received recent attention from a number of researchers [4,20–22].

In the relevant literature, piezoelectric signals have usually been measured by charge amplifiers, which detected the polarized charges on bone surfaces by transferring them to a capacitor in an instrument. This is a relatively simple way to determine the quantitative relationships between the electric charge and external force. However, this method cannot reveal the time-dependent variation of the polarized charges in bone, during which the magnitude of charges may influence the bone remodeling process or osteocyte viability. Over the past decades, various new techniques have been developed and employed to investigate the piezoelectric properties of bone tissue. Based on the concept of converse piezoelectric effect, the piezoelectric coefficient d_{23} has been determined using a sensitive dilatometer [20]. A piezoelectric force microscope and an atomic force microscope have also been employed to measure piezoelectric properties of bone [21,22]. More recently, Hou, Fu, and Qin [4] investigated the stretching–relaxation properties of bone piezo-voltages. Fu et al. [23] studied the influence of shear stress on the signs of piezovoltages in bone by means of three-point and four-point bending experiments. In this chapter, our discussion focuses on the developments in references 3–5 and 23.

9.2 Removal of Soft Tissue from Bone Samples

Cleaning of fresh bones to remove their soft tissues while preserving their structural integrity is a basic, practical, and essential part of bone studies. This section briefly summarizes the development presented in Yin et al. [5]. It describes the effects of water, trypsin, and detergent macerations on bone microstructures and mechanical properties during the preparation process. In particular, two typical laundry powders—one without enzymes and the other with two enzymes—are used.

9.2.1 Removal of Soft Tissues

In Yin et al. [5], three maceration fluids containing different additives and 2 L of water are evaluated (Table 9.1) and compared with the same amount of additive-free water. Solution A is additive-free water. Solution B contains 20 g trypsin (Sigma Aldrich, Australia). The molecular weight of trypsin is 23,800, and it consists of 223 amino acids. The pH value of trypsin is 7.8. Solution C contains 40 g Surf laundry powder (Unilever, Australia). There is no added enzyme in Surf laundry powder. The pH value of Solution C is 11. Solution D contains 40 g Biozet laundry powder (KAO, Australia).

TABLE 9.1

Test Solutions

Solution	Additive	pH Value
A	2 L Water only	7
B	20 g Trypsin powder in 2 L water	7.8
C	40 g Surf powder in 2 L water	11
D	40 g Biozet powder in 2 L water	10.5

Biozet laundry powder contains two types of enzymes for biologically active cleaning, anionic and nonionic surfactant for lifting dirt from clothes, sodium perborate monohydrate for oxygen bleach, sodium alumino silicate for softening water, sodium carbonate for breaking up fatty soils, fluorescers for brightening fabric, soil-suspending agent, and perfume (KAO, Australia). The pH value of Solution D is 10.5.

In their experiment Yin et al. [5] used 16 lamb femurs from 6-month old lambs purchased from a local market, which were stored in a refrigerator at –20°C before all the joints were cut off using a diamond saw machine. Four femurs were macerated in each solution and then stored in a closed plastic container for 5 days at room temperature in a fume cupboard. After 5 days, manual removal of soft tissues was conducted using a rod, a cooking knife, and a brush. Care was taken to avoid scraping, scratching, or cutting bone surfaces. For safety reasons, all the manual removal was conducted in the fume cupboard. When evaluating maceration of bones for soft tissue cleaning, several aspects must be considered, including odor, soft-tissue texture, ease of flesh removal, and bone quality [6].

The cleaned bone samples were stored in an isotonic phosphate buffered saline (PBS) solution, with sodium azide as preservative, at room temperature. This solution, containing the identical mineral content to mammalian cells, was prepared by dissolving 800 g NaCl, 20 g KCl, 144 g Na_2HPO_4, 24 g KH_2PO_4, and 0.2% NaN_3 in 8 L of distilled water, and topping up to 10 L. The pH of the solution was about 6.8.

9.2.2 Preparation of Thin Sections

Yin et al. processed each femur macerated in each solution for thin section preparation for microstructural analysis by scanning electron microscopy following a standard protocol for thin section preparation.

Using a diamond saw, they then cut a suitably sized slab from the center part of each femur macerated in each solution for mounting on a slide. The slab was labeled on one side and the other side was lapped flat and smoothed first on a cast iron lap with 400 grit carborundum, then finished on a glass plate with 600 grit carborondum. After the slab had been dried on a hot plate, a glass slide was glued to the lapped face of the slab with epoxy. Using a thin section saw, the slab was cut off close to the slide. The thickness was further reduced on a thin section grinder. A finished thickness of 30 µm was achieved by lapping the section by hand on a glass plate with 600 grit carborundum. Then, fine grinding with 1,000 grit was conducted. After that, the section was placed in a holder and spun on a polishing machine using nylon cloth and diamond paste until a polish suitable for microscope study was achieved.

9.2.3 Microstructural Analysis and Porosity Measurement

Yin et al. [5] carbon-coated the thin sections described previously and observed their microstructures under a scanning electron microscope (Cambridge 360, Cambridge, UK). They took back-scattered images under high vacuum at 20 kV. The regions of interest of the scanning electron micrograph (SEM) images were transferred into commercial image processing software (Analysis X) to perform pore analysis. Each pore area was labeled and calculated and the maximum and mean pore sizes were obtained in regions of interest of 106,875 µm². The porosity expressed in percent was defined by the ratio of the total area of the pores to the total areas of the regions of interest. On each thin section obtained from each femur macerated from each solution, three random locations were measured to obtain the mean value and the standard deviation. Analysis of variance (ANOVA) at a 5% significance level was applied for the statistical analyses of porosities.

9.2.4 Standard Microhardness Indentation Testing

Using a diamond saw machine at a low rotary speed, Yin et al. [5] then obtained transverse sections 10 mm in thickness from the central femurs macerated in the four solutions. During the cutting process, alcohol was applied as coolant. The 10 mm section specimens were cleaned to remove any abrasives from the diamond cutting and preserved in PBS solutions. The transversely cut specimens were polished using metallographic polishing techniques. The initial polishing was performed on a series of silicon carbide grit papers with sizes of 240, 320, 400, 600, 800, and 1,200 grit. The fine polishing was performed using diamond suspension slurries with grades of 6, 3, 1, and 0.25 µm on polishing cloth. The specimens were cleaned before proceeding to the next finer level of polishing. After the final polishing, the bone samples were stored in the PBS solution at room temperature.

The polished bone surfaces were indented with a Vickers diamond indenter in a standard microhardness tester (MHT-1, Matsuzawa Seiki, Japan). Five indentation loads of 0.245, 0.49, 1.96, 4.9, and 9.8 N were applied for the loading time of 10 s each. Six indentations were made for each load on each sample. This resulted in a total of 30 indentations in each sample. A distance of at least twice the impression diagonal was maintained between the indentations to minimize interactions between neighboring indentations. The indentations were completed within 45 min from the time each bone sample was taken out of the PBS solution. The indentation diagonals were measured with optical microscopy. Three samples from the same maceration were selected for repeat tests. ANOVA at a 5% significance level was applied for the statistical analysis of hardness values.

9.2.5 Results for the Samples after Removal of Soft Tissues

Yin et al. [5] found that all the bone samples showed appreciable change after the 5-day maceration in each solution at room temperature. All the marrows were easily removed using a wooden rod and a brush. The soft tissues of the bone samples macerated in solution A (water) were difficult to remove. Cleaning was conducted carefully by scratching the soft tissues using a knife. The bone samples in solution B (trypsin) become completely disarticulated and the soft tissues could be easily separated and cleaned with a scouring pad without any damage to bone surfaces. In comparison of the dissolving abilities between the two laundry detergents, Surf (solution C) and Biozet (solution D), the bone samples macerated in Surf still showed a large amount of soft tissue adhering on the bone surfaces, whereas the samples treated with Biozet were almost free of soft tissue, and remnants of soft tissue could be easily removed.

A noxious chemical odor was associated with the samples macerated in water and trypsin. In particular, the bone samples treated in trypsin were extremely malodorous at room temperature, causing strong offense to people. The smells of the bone samples treated with Biozet and Surf laundry powders were much more acceptable. In particular, the bone samples treated with Biozet produced a pleasant odor.

9.2.6 Change of Microstructure with Cleaning Procedure

The SEM micrographs demonstrating the microstructures of the bone samples prepared in each solution are shown in Figure 9.1 [5]. There are no visible differences among these four microstructures. All microstructures show the typical osteonal structure of compact (cortical) bone. Figure 9.2 shows typical osteons, which were formed rather like plywood from sheets of alternating lamellae that could be laid flat, or curved around in circles to protect blood vessels. The osteons took the form of planar ellipses. One had a major axis approximately 100 µm in length and a minor axis approximately 50 µm in length. The pores with diameters larger than 10 µm were formed from Haversian canals or blood vessels. Smaller pores were formed from lymphatic vessels and lacuna canaliculi.

Figure 9.3(a) demonstrates one example of identification, analysis, and calculation of porosity in a region of interest. Figure 9.3(b) shows the pore size distribution in the region, indicating that 4% of pores have a diameter of 10–20 µm and 96% of pores have a diameter smaller than 10 µm. The porosities

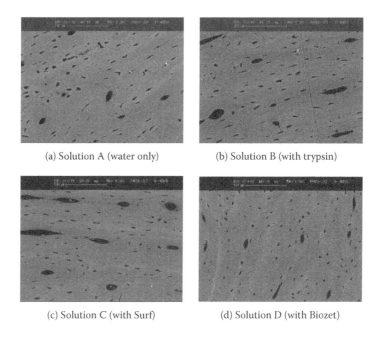

(a) Solution A (water only)   (b) Solution B (with trypsin)

(c) Solution C (with Surf)   (d) Solution D (with Biozet)

FIGURE 9.1
Scanning electron micrographs of the microstructures of the bones macerated in the four solutions.

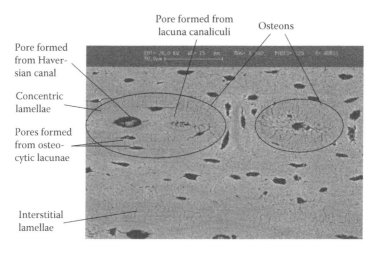

FIGURE 9.2
SEM micrograph of the microstructure of the compact bone.

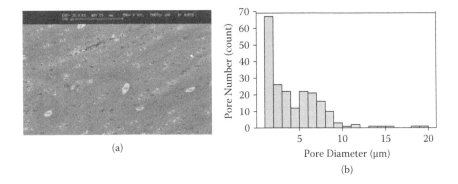

(a)

(b)

FIGURE 9.3
(a) Porosity identification and (b) distribution of pore diameters.

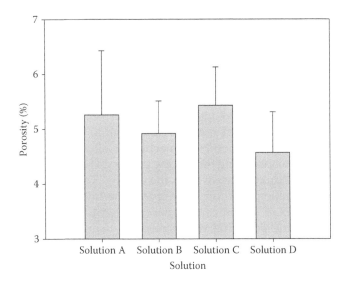

FIGURE 9.4
Porosity of bone macerated in each solution. Each point is the average value from three repeated measurements and the error bar is ±one standard deviation of the repeated measurements.

measured for the bones prepared under different macerations are plotted in Figure 9.4. Each datum is the average with one standard deviation of three measurements at random locations of the bone sample. The mean porosities of bones macerated in water, trypsin, Surf, and Biozet were 5.26%, 4.92%, 5.34%, and 4.57%, respectively. Statistical analysis indicated that there was no significant difference among the porosities of bones macerated in the different solutions (ANOVA, $p > 0.05$).

9.3 Microindentation Testing of Dry Cortical Bone Tissues

In Section 9.2, approaches for removing the soft tissue from bone samples were described. Microindentation testing of these cleaned bone samples as presented in Yin et al. [3] is discussed in this section. Typically, assessments of the crack propagation and deformation behavior of dry cortical bone on a microscopic scale by means of microindentation are described. It is well known that microindentation offers a means of characterizing the deformation properties of bone [24] and provides a basis for evaluating a range of contact-related properties, particularly surface-damage phenomena in sharp-particle compression [25].

9.3.1 Preparation of Bone Samples

To conduct the microindentation testing described in this section, Yin et al. [3] purchased three lamb femurs from 6-month-old lambs from a local supermarket and stored these femurs in a refrigerator at –20°C. The femurs were macerated in a solution containing 40 g Biozet laundry powder (KAO, Australia) and 2 L water for 5 days at room temperature in a fume cupboard. (The composition of Biozet was described in Section 9.2.1.) The pH value of the solution was 10.5. After 5 days, soft tissues were manually removed from the femurs using a rod, a cooking knife, and a brush.

Yin et al. then cut transverse section samples of 10 mm thickness from the central femurs using a diamond saw machine at a low rotary speed. Alcohol was utilized as coolant during the cutting process. The samples were washed to remove any residual abrasive from the cutting. They were then polished using metallographic polishing techniques. The initial polishing was performed on a series of silicon carbide papers of grit sizes 240, 320, 400, 600, 800, and 1,200. Fine polishing was performed using diamond suspension slurries with grades 6, 3, 1, and 0.25 μm on polishing cloth. The samples were cleaned before proceeding to the next finer level of polishing. After the final polishing, the samples were stored in a fume cupboard at room temperature for 2 weeks.

9.3.2 Standard Microhardness Indentation Testing

The polished dry bone surfaces were indented with a Vickers diamond indenter in a microhardness tester (MHT-1, Matsuzawa Seiki, Japan) [3]. Five indentation loads of 0.245, 0.49, 1.96, 4.9, and 9.8 N were applied for 10 s. Six indentations were made at each load on each transverse section. This resulted in a total of 30 indentations in each sample. A distance of at least twice the impression diagonal was kept between indentations to prevent interactions between neighboring indentations. The lengths of the indentation diagonals were measured using optical microscopy. Three samples were selected for repeat tests and production of a total of 18 indentations at each load. The indented samples were carbon coated and stored in an oven at 42°C for 48 h so that they could be easily vacuumized in the SEM sample chamber. Samples were observed under a SEM (Cambridge 360, Cambridge, UK). Backscattered electron (BSE) imaging was performed at 20 kV accelerating voltage at a working distance of 16 mm, at 350× and higher magnifications. Both the optical microscope and the SEM were calibrated prior to the length measurements and had relative system uncertainties of about 2%.

9.3.3 Testing Results

Yin et al. [3] reported their results obtained from the microindentation experiments they conducted in Figures 9.5–9.8. Figure 9.5 shows Vickers hardness against applied load. Each datum is the average with one standard deviation of 18 indentations in three samples. The figure demonstrates that the hardness increased with the applied load. In particular, the hardness significantly increased with the load at the low loads of 0.245 and 0.49 N (ANOVA, $p < 0.05$). After the load reached 1.96 N, hardness was independent of the load (ANOVA, $p > 0.05$). With an increase in the load from 0.245 to 4.9 N, hardness increased by 15% from a mean of 0.45 to 0.54 GPa, but hardness values remained approximately unchanged when the load was increased from 1.96 to 9.8 N.

Figure 9.6 shows a series of BSE images of indentation patterns at the applied loads of 9.8, 4.9, and 1.96 N at 350× magnification. No cracks were clearly observed in the indentation patterns for any applied load. Plastic deformation was observed in all the indented areas. However, by measuring the

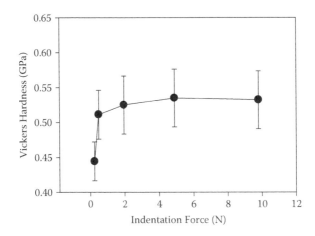

FIGURE 9.5
Vickers hardness of dry bone (lamb femurs). Each point is the average with the error bar of ±one standard deviation of 18 indentations in three samples.

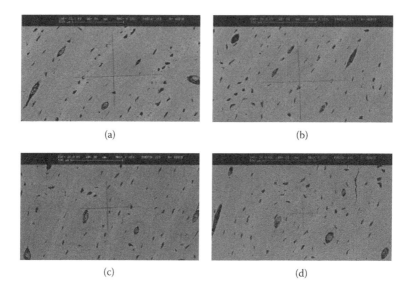

(a)   (b)

(c)   (d)

FIGURE 9.6
BSE images of the indentation patterns at loads (a) and (b) 9.8 N, (c) 4.9 N, and (d) 1.96 N.

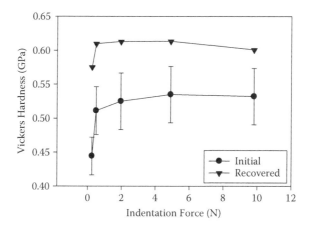

FIGURE 9.7
Recovered Vickers hardness measured under a SEM compared with the initial measurement shown in Figure 9.5. Note that the recovered hardness datum is a single measurement with the SEM system error less 2%.

indentation diagonals under the SEM, the indentation patterns were found to be apparently smaller than the initial hardness indentation measured using optical microscopy immediately after the indentations were made.

Figure 9.7 shows the recovered Vickers hardness values calculated from SEM measurements compared with the initial values measured immediately

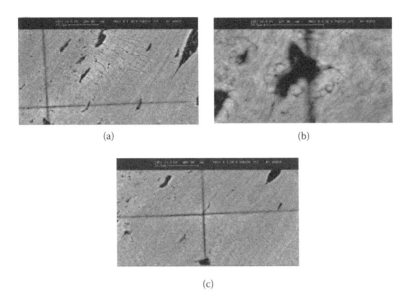

(a)　　　　　　　　　　　　(b)

(c)

FIGURE 9.8
High-magnification BSE images of the indentation patterns at load 9.8 N, showing (a) a micro-crack propagating from a large Haversian canal with a length of approximately 10 μm and microcrack clusters developing from the boundaries of small osteocyte lacunae with lengths less than 20 μm; (b) microcracks formed in the woven fibrils from a canaliculus, with lengths less than 5 μm; and (c) a microcrack developing at the apex of indentation along the longitudinal lamellae.

after indentation in the hardness tester. Each recovered hardness datum is a single measurement with the SEM system error less than 2%; the initial hardness datum is the average with one standard deviation of 18 indentations. The hardness values calculated after recovery were found to be not only significantly higher than the initial hardness mean value (ANOVA, $p < 0.05$) but also significantly higher than the initial hardness mean value plus one standard deviation (ANOVA, $p < 0.05$). In particular, at least 12% recovery of the indentation deformation occurred at loads lower than 0.45 N, whereas at least 5% recovery of the deformed bone occurred at loads higher than 1.96 N.

Figure 9.6(a) and 9.6(b) show that at load 9.8 N, the indented area covers several osteons, woven fibrils, and interstitial lamellae [3]. To reveal deformation details of the indentations, high-magnification BSE images are shown in Figure 9.8. Figure 9.8(a) demonstrates a microcrack propagated from a large Haversian canal with a length of approximately 10 μm. Yin et al. [3] also indicated that microcrack clusters developed from the boundaries of small osteocyte lacunae in the bone with lengths smaller than 20 μm. Figure 9.8(b) shows microcracks formed in the woven fibrils from a canaliculus, with lengths smaller than 5 μm. Figure 9.8(c) reveals a microcrack developing at the apex of an indentation along the longitudinal lamellae.

Another direct observation made by Yin et al. [3] was the significant recovery of indentation deformation as shown in Figure 9.7. In particular, at lower applied loads than 0.45 N, the bone exhibited more viscoelasticity than at high loads. Bone recovery following microindentation has also been observed in wet bovine femur [26]. The time dependence of viscoelasticity in bone has been observed in R-curve testing of human cortical bone as well [13]. Time-dependent crack growth occurs in bone under sustained (noncyclic) in vitro loads at stress intensities lower than the nominal crack-initiation toughness [13]. Effects of viscoelasticity and time-dependent plasticity of human cortical bone have been further investigated using nanoindentation [27]. However, the exact nature of such behavior is as yet unclear [13]. Hardness has been found to increase with load, as shown in Figure 9.5. This phenomenon has also been observed in microindentation of dry embalmed human rib [28]. However, in microindentation of wet bovine metacarpus, hardness decreased with the applied load [16].

9.4 Stretching–Relaxation Properties of Bone Piezovoltage

In Sections 9.2 and 9.3, experiments on the mechanical behavior of bones were discussed. Experiments on the coupling behaviors of mechanical and electric fields (e.g., piezoelectric behavior) in bone are described in this and the subsequent section. Experiments presented in Hou et al. [4] on stretching–relaxation properties of bone piezovoltage are described in this section. These researchers measured piezovoltages between the two opposing surfaces of bovine tibia bone samples under three-point bending deformation using an ultrahigh-input impedance bioamplifier. The experimental results presented there indicated that the piezovoltages of bone showed different relaxation behaviors during loading and unloading processes. Hou et al. found that the piezovoltage decay followed a stretched exponential law when the load increased from zero to its maximum value, whereas it followed a typical relaxation exponential law when the load was maintained at its maximum value. The stretching-exponential behavior was independent of loading amplitude and rate.

9.4.1 Sample Preparation

As indicated in Hou et al. [4], the cortical bone samples used were harvested from mid-diaphysis of dry bovine tibias (age 2–3 years) and machined into rectangular beams (Figure 9.9a) with the dimension range shown in Table 9.2. Six samples were prepared and dried in air for at least 2 weeks until the resistance or impedance between the two lateral surfaces was over 10^9 Ω, and the maximum resistance of complete dry bone is usually in the order

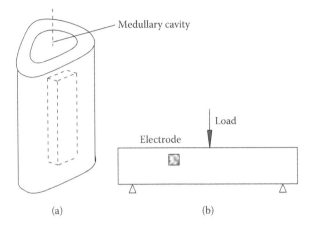

FIGURE 9.9
(a) The geometry of the sample; (b) three-point test.

TABLE 9.2

Size of Samples

Span (mm)	Width (mm)	Height (mm)
85	4.0 ± 1.0	19 ± 3.0

of 10^{10} Ω [29]. Then, the conductive silver adhesive (5001, SPI, United States) was painted on both sides of each specimen as electrodes with dimensions of 3 mm × 3 mm (Figure 9.9b). After the electrodes was painted, the samples were kept in an environment with relative humidity of 52%–56% and temperature of 22°C–25°C. The two electrodes were placed in the same axial position and at the same mid-height from the neutral axes of the samples, where the bone samples are subjected to both normal stress and shear stress.

9.4.2 Experimental Setup

The principle of the experimental setup was as follows [4]. It was noted that the pair of electrodes and the bone tissue sandwiched between them were equivalent to a capacitor; C was defined as its capacitance [29,30]. Once the piezoelectric charge Q was induced by the load accumulating on the electrodes, a corresponding piezovoltage V (V = CQ) could be observed. Because piezovoltage is a linear function of piezocharge, the variation of the voltage is consistent with that of the charge. Based on this principle, Hou et al. designed the experimental setup of the measurement system [4] as illustrated in Figure 9.10. It can be seen from this figure that the electrode on the lateral side facing the medullary cavity (Figure 9.9) was taken as zero potential (or reference potential). The measured piezovoltage was input into a bioamplifier

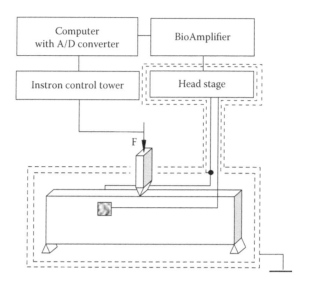

FIGURE 9.10
Setup of the test system.

(BMA-931, CWE Inc., United States) via an ultrahigh-input impedance (over 10^{12} Ω) head stage (Super Z, CWE Inc., United States), whose input impedance was at least two orders higher than that of bone and was able to prevent the charges accumulated on the electrodes from leaking through the head stage. In the measurement, the amplified voltage signals were recorded by a computer. Loads were applied using an Instron 1343 closed-loop servo-hydraulic machine controlled by an 8800 Control Tower, and the loading signals in the control tower were also input into the computer for recording.

The testing sample and the head stage were enclosed in a double electromagnetic shield box with the outer shield connecting to ground and the inner to the head stage common terminal. This arrangement kept the electric field distribution constant in the device [31].

9.4.3 Experimental Procedure and Characteristics of Piezovoltage

Figure 9.11 shows a trapezoidal loading configuration applying to the samples in the experiment reported in Hou et al. [4]. It has equal loading and unloading time t_o. (These researchers employed $t_o = 0.25, 0.5$, and 1 s.) Having reached its maximum F_o ($F_o = 50, 100$, and 150 N in the experiment), the load was kept constant for 6 s and then decreased to zero (see Figure 9.11a). Hou et al. noted that the cortical bone had weak viscoelasticity, especially dry bone [32]. To reduce the effect of viscoelasticity, the maximum compressive stresses in the samples, caused by the maximum F_o, were between 12 and 23 MPa, which was much less than 80 Mpa, below which no irrecoverable

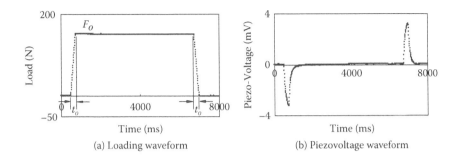

FIGURE 9.11
Loading and piezo-voltage wave form.

deformation occurs in cortical bone [33] and stress and strain have a linear relationship for cortical bovine bone [34].

In Hou et al. [4], piezovoltages of the bone samples were measured under three-point bending in an environment with relative humidity of 52%–56% and temperature of 21°C–26°C. Figure 9.11(b) shows a typical plot of variation of piezovoltage with time under the loading profile shown in Figure 9.11(a). It has a negative and a positive pulse corresponding to the loading and unloading durations, respectively. The peak of the first pulse (negative) just corresponds to the loading end point and the second pulse peak corresponds to the unloading end point. The amplitudes of the peaks are of the order of several millivolts. The pattern of the pulses indicates that once the loading or unloading ends, the piezovoltage starts to decay toward zero, which looks like an exponential relaxation process. The corresponding physical process was that once the piezocharge appeared on the two electrodes, it began to discharge through the impedance of the bone. Then, the variation of the piezocharge with time was associated with the mechanical loads and with the physical properties of the bone, such as impedance.

9.4.4 Results and Discussion

Hou et al. [4] found that the piezovoltage time curves of all the samples (Figure 9.12) were similar in shape under three-point bending. Figures 9.12 and 9.13 illustrate two groups of these curves for samples numbered 1 and 2, respectively, with different loading conditions.

After many trials of curve fitting on the measured curves (details of the fitting scheme are described in Section 9.4.5), Hou et al. found that the piezovoltages showed different relaxation behaviors in the loading (or unloading) and load holding processes. During loading, the piezovoltage decays followed a nonexponential or stretched exponential law [35], whereas they followed a typical exponential law ($\beta = 1$, known as Debye exponential

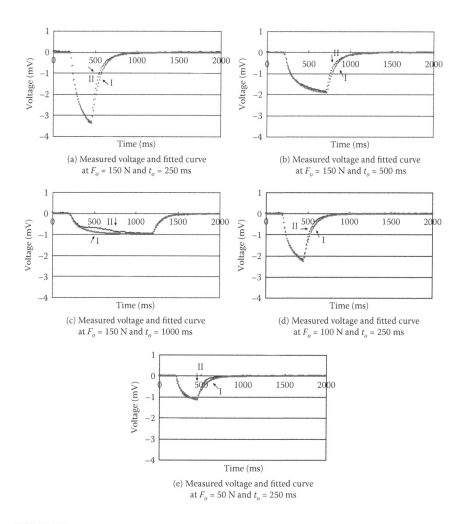

FIGURE 9.12
Fitting functions with different F_o and t_o for sample no. 1 (I: fitting results; II: measured results).

relaxation) during load holding. Equation (9.1) represents the fitting function for the measured piezovoltage $V(t)$:

$$V(t) = KF(t)e^{-\left(\frac{t}{\tau_d}\right)^{\beta}} \tag{9.1}$$

where
  t is time
  $F(t)$ is the load function
  K is a proportional coefficient between load $F(t)$ and piezovoltage

τ_d is time constant or relaxation time

$\beta(0 < \beta \leq 1)$ is a stretching exponent

In the experiment presented in Hou et al. [4], the first stage of the loading process $F(t)$ was given as

$$F(t) = F_0 \frac{t}{t_0} \quad (t \leq t_0) \tag{9.2}$$

Equation (9.1) indicates that the piezovoltage is generated in proportion to the load $F(t)$ and it decays in the stretched exponential law simultaneously.

When the loading remains constant, the fitted piezovoltage function is

$$V(t) = V_o e^{-\frac{t}{\tau_c}} \tag{9.3}$$

where τ_c is a time constant and V_o is the peak voltage when the first stage of the loading process ends. Letting $t = t_o$ and substituting Equation (9.2) into Equation (9.1) yields

$$V_o = K \cdot F_o \cdot e^{-\left(\frac{t_o}{\tau_d}\right)^{\beta}} \tag{9.4}$$

Equation (9.3) represents that the piezovoltage decays in a typical exponential law. The fitted functions for the piezovoltage of sample 1 are then written as

$$V(t) = \begin{cases} -0.271 F_o \dfrac{t}{t_o} e^{-\left(\frac{t}{11.79}\right)^{0.2993}} & (t \leq t_o) \\[4mm] -0.271 F_o e^{-\left(\frac{t_o}{11.79}\right)^{0.2993}} \cdot e^{-\left(\frac{t}{94}\right)} & (t > t_o) \end{cases} \tag{9.5}$$

In Figure 9.12(a)–(e) the curves represent the fitting function (9.5) with different F_o and t_o and the curves are the corresponding measured piezovoltages. Figure 9.12(a)–(c) shows three fitting curves with $F_o = 150$ N and $t_o = 250$, 500, and 1000 ms, respectively. Figure 9.12(d) and (e) shows two fitting curves with $t_o = 250$ ms and $F_o = 50,100$ N, respectively. It is evident that the fitted functions coincide well with the measured curves. The values of K, τ_d, β, and τ_c are listed in Table 9.3.

The fitted functions for sample 2 are as follows [4]:

$$V(t) = \begin{cases} -0.295 F_o \dfrac{t}{t_o} e^{-\left(\frac{t}{12.74}\right)^{0.302}} & (t \leq t_o) \\[4mm] -0.295 F_o e^{-\left(\frac{t_o}{12.74}\right)^{0.302}} \cdot e^{-\left(\frac{t}{93}\right)} & (t > t_o) \end{cases} \tag{9.6}$$

Again, in Figure 9.13(a)–(f) the curves I represent the fitting function (9.6) with different F_o and t_o and the curves II are the corresponding measured piezovoltages. Figure 9.13(a)–(c) shows three fitting curves with F_o = 150 N and t_o = 250, 500, and 1000 ms, respectively; Figure 9.13(d)–(f) shows the same three measured curves but corresponding to the unloading portion, in which the red curves are obtained using the same fitting functions as those in Figure 9.13(a)–(c). The coincidences of the curves in Figure 9.13(d)–(f) imply that the piezovoltage curves during the unloading process have similar waveforms to those in the loading process.

Hou et al. [4] found that all the piezovoltage curves of the six samples had the same form of fitting function. The first part (stretched exponential term)

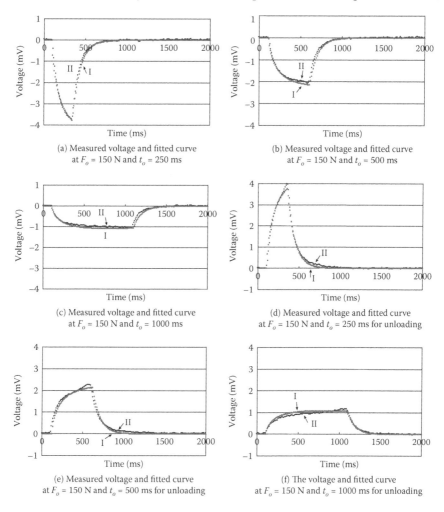

(a) Measured voltage and fitted curve
at F_o = 150 N and t_o = 250 ms

(b) Measured voltage and fitted curve
at F_o = 150 N and t_o = 500 ms

(c) Measured voltage and fitted curve
at F_o = 150 N and t_o = 1000 ms

(d) Measured voltage and fitted curve
at F_o = 150 N and t_o = 250 ms for unloading

(e) Measured voltage and fitted curve
at F_o = 150 N and t_o = 500 ms for unloading

(f) The voltage and fitted curve
at F_o = 150 N and t_o = 1000 ms for unloading

FIGURE 9.13
Fitting functions at different F_o and t_o for sample no. 2 (I: fitting results; II: measured results).

TABLE 9.3

Fitting Parameters

Sample no.	Fitting Parameters			
	β	K	τ_d (ms)	τ_c (ms)
1	0.299	−0.272	11.79	94
2	0.302	−0.295	12.74	93
3	0.305	−0.212	10.98	60
4	0.276	−0.190	11.44	115
5	0.286	−0.214	9.30	81
6	0.291	−0.213	11.42	88
Mean ± StDev	0.293 ± 0.01	−0.233 ± 0.038	11.28 ± 1.037	88.5 ± 16.4

of the fitting function is characterized by three parameters: the proportional coefficient K, stretching exponent β, and time constant τ_d. The second part (typical exponential term) is characterized by only one parameter: the time constant τ_c, which is easily determined. These parameters are independent of the holding load F_o and loading time t_o. To determine the three parameters in the first part, at least five piezovoltage time curves were measured for each sample corresponding to different holding loads F_o and loading times t_o, as shown in Figure 9.12. Hou et al. obtained the three parameters by fitting any three of the five curves to Equation (9.1) using a trial-and-error method on a computer. Then the other two curves were fitted by the three determined parameters, with corresponding values of F_o and t_o, to check the adequacy of fitting. Hou et al. found that once the three parameters were determined by any three curves, they were available for the remaining two curves. This proved the uniqueness of the fitting. In other words, the stretched exponential behavior merely represents the bone's inherent property. The fitting parameters of the six samples are listed in Table 9.3.

The time constant τ_d in the loading process was about one order of magnitude smaller than the τ_c in the loading hold process. The significance of the stretched exponent $\beta(0.276 < \beta < 0.305)$ is that the relaxation mechanism in the loading process was different from that in the load holding process. The fitting functions (9.5) and (9.6) and the measured piezovoltage curves in Figures 9.12 and 9.13 show that the peaks of piezovoltage are proportional to the maximum load F_o and inversely proportional to loading interval t_o. Making use of the mean value of the fitting perameters listed in Table 9.3, the fitting function (9.1) of the bone can be further written as

$$V(t) = \begin{cases} -0.233 F_o \dfrac{t}{t_o} e^{-\left(\frac{t}{11.28}\right)^{0.293}} & (t \leq t_o) \\[4mm] -0.233 F_o e^{-\left(\frac{t_o}{11.28}\right)^{0.293}} \cdot e^{-\left(\frac{t}{88.5}\right)} & (t > t_o) \end{cases} \tag{9.7}$$

9.4.5 The Fitting Scheme for Stretched Exponential Function

Substituting Equation (9.2) into Equation (9.1) yields

$$V(t) = KF_0 \frac{t}{T_o}(t)e^{-\left(\frac{t}{\tau_d}\right)^\beta} \qquad t \leq T_0 \tag{9.8}$$

As there are three unknown parameters, K, τ_d, and β, in Equation (9.8), three measured piezovoltage time curves at different load magnitudes, F_o, and loading times, T_o, of each sample were employed to determine them.

For the convenience of calculation and without loss of generality, Hou et al. [4] used the measured piezovoltage time curves, denoted as $\bar{V}(t)$, at loading time $T_o = 0.25$, 0.5, and 1 s, respectively, and load magnitude $F_o = 100$ N in the fitting calculation. For simplicity, denote the loading time $T_o = 0.25$, 0.5, and 1 s and the corresponding three measured piezovoltages as T_{025}, T_{05}, and T_1 and $\bar{V}_{025}(t)$, $\bar{V}_{05}(t)$, and $\bar{V}_1(t)$, respectively. Further, denote the time t at 0.25, 0.5, and 1 s as t_{025}, t_{05}, and t_1. Substituting the three piezovoltages into Equation (9.8), respectively, Hou et al. [4] obtained the following three equations:

$$\bar{V}_{025}(t_{025}) = KF_0 \frac{t_{025}}{T_{025}}e^{-\left(\frac{t_{025}}{\tau_d}\right)^\beta} \qquad t_{025} \leq T_{025} \tag{9.9}$$

$$\bar{V}_{05}(t_{05}) = KF_0 \frac{t_{05}}{T_{05}}e^{-\left(\frac{t_{05}}{\tau_d}\right)^\beta} \qquad t_{05} \leq T_{05} \tag{9.10}$$

$$\bar{V}_1(t_1) = KF_0 \frac{t_1}{T_1}e^{-\left(\frac{t_1}{\tau_d}\right)^\beta} \qquad t_1 \leq T_1 \tag{9.11}$$

The ratio of Equation (9.9) to Equation (9.10) yields

$$\frac{\bar{V}_{025}(t_{025})}{\bar{V}_{05}(t_{05})} = \frac{T_{05}t_{025}}{T_{025}t_{05}}e^{\left(\frac{t_{05}}{\tau_d}\right)^\beta - \left(\frac{t_{025}}{\tau_d}\right)^\beta} \tag{9.12}$$

Applying the logarithm to Equation (9.12) leads to

$$\ln\left[\frac{\bar{V}_{025}(t_{025})}{\bar{V}_{05}(t_{05})}\frac{T_{025}t_{05}}{T_{05}t_{025}}\right] = \frac{t_{05}^\beta - t_{025}^\beta}{\tau_d^\beta} \tag{9.13}$$

Thus, the time constant τ_d equals

$$\tau_d = \left\{ \frac{t_{05}^{\beta} - t_{025}^{\beta}}{\ln\left[\dfrac{\bar{V}_{025}(t_{025})}{\bar{V}_{05}(t_{05})} \dfrac{T_{025}t_{05}}{T_{05}t_{025}}\right]} \right\}^{\frac{1}{\beta}} \tag{9.14}$$

By Equation (9.8), the proportional coefficient K is

$$K = \frac{\bar{V}_{025}(t_{025})}{F_0 \dfrac{t_{025}}{T_{025}} e^{-\left(\frac{t_{025}}{\tau_d}\right)^{\beta}}} \tag{9.15}$$

It is clear that once the stretching exponent β is known, τ_d and K are obtained. Moreover, similar equations to Equations (9.14) and (9.15) can be obtained from the ratios of the other equations. To distinguish the parameters calculated from other ratios, τ_d and K in the preceding equations are replaced by τ_{d12} and K_{12}. Thus, Equations (9.14) and (9.15) can be rewritten as

$$\tau_{d12} = \left\{ \frac{t_{05}^{\beta} - t_{025}^{\beta}}{\ln\left[\dfrac{\bar{V}_{025}(t_{025})}{\bar{V}_{05}(t_{05})} \dfrac{T_{025}t_{05}}{T_{05}t_{025}}\right]} \right\}^{\frac{1}{\beta}} \tag{9.16}$$

$$K_{12} = \frac{\bar{V}_{025}(t_{025})}{F_0 \dfrac{t_{025}}{T_{025}} e^{-\left(\frac{t_{025}}{\tau_{d12}}\right)^{\beta}}} \tag{9.17}$$

Similarly, τ_{d23}, K_{23}, τ_{d31}, and K_{31} can be defined as

$$\tau_{d23} = \left\{ \frac{t_{1}^{\beta} - t_{05}^{\beta}}{\ln\left[\dfrac{\bar{V}_{05}(t_{05})}{\bar{V}_{1}(t_{1})} \dfrac{T_{05}t_{1}}{T_{1}t_{05}}\right]} \right\}^{\frac{1}{\beta}} \tag{9.18}$$

$$K_{23} = \frac{\bar{V}_{05}(t_{05})}{F_0 \dfrac{t_{05}}{T_{05}} e^{-\left(\frac{t_{05}}{\tau_{d23}}\right)^{\beta}}} \tag{9.19}$$

$$\tau_{d31} = \left\{ \frac{t_{025}^{\beta} - t_1^{\beta}}{\ln\left[\dfrac{\bar{V}_1(t_1)}{\bar{V}_{025}(t_{025})} \dfrac{T_1 t_{025}}{T_{025} t_1} \right]} \right\}^{\frac{1}{\beta}}$$

(9.20)

$$K_{31} = \frac{\bar{V}_1(t_1)}{F_0 \dfrac{t_1}{T_1} e^{\left(\frac{t_1}{\tau_{d1}}\right)^{\beta}}}$$

(9.21)

Hou et al. [4] then indicated that the ideal fitting result is to find a specific value of β that makes

$$\tau_d = \tau_{d12} = \tau_{d23} = \tau_{d31}$$

and

$$K = K_{12} = K_{23} = K_{31}$$

Due to the presence of measurement errors, the fitting criterion is used to find a $\beta(0 < \beta < 1)$ which minimizes the error function

$$\text{Error} = \left(|\tau_{d12} - \tau_{d23}| + |\tau_{d23} - \tau_{d31}| + |\tau_{d31} - \tau_{d12}| \right)$$
$$\times \left(|K_{12} - K_{23}| + |K_{23} - K_{31}| + |K_{31} - K_{12}| \right)$$

(9.22)

under the six restrictive conditions $|\tau_{d12} - \tau_{d23}| \leq \Delta\tau$, $|\tau_{d23} - \tau_{d31}| \leq \Delta\tau$, $|\tau_{d31} - \tau_{d12}| \leq \Delta\tau$, $|K_{12} - K_{23}| \leq \Delta K$, $|K_{23} - K_{31}| \leq \Delta K$, and $|K_{31} - K_{12}| \leq \Delta K$, where $\Delta\tau$ and ΔK are error thresholds of time constants and proportional coefficients, respectively. Then the values of τ_d and K are defined as their average value: $\tau_d = (\tau_{d21} + \tau_{d23} + \tau_{d31})/3$ and $K = (K_{12} + K_{23} + K_{31})/3$. Because τ_d and K are not of the same order of magnitude, the error function is configured by multiplying the sum of the three time constant differences and the sum of the three proportional coefficient differences.

Equations (9.16)–(9.21) imply that if a specific β is found K, τ_d can be determined based on the three piezovoltage values of $\bar{V}_{025}(t_{025})$, $\bar{V}_{05}(t_{05})$, and $\bar{V}_1(t_1)$ at a specific time t. For the curve fitting, the three piezovoltages of $\bar{V}_{025}(t_{025})$, $\bar{V}_{05}(t_{05})$, and $\bar{V}_1(t_1)$ at $t_{025} = 0.25$ s, $t_{05} = 0.5$ s, and $t_1 = 1$ s, respectively, are used, which correspond to the three peak values of the three piezovoltage time curves.

To find suitable fitting results under this criterion, Hou et al. [4] carried out circular calculations for the preceding equations when β increased from 0 to 1 step by step. The parameters K, τ_d, and β obtained are listed in Table 9.3.

Hou et al. [4] found from their experimental results that if the value of $\Delta\tau$ or ΔK was set to be less than 5%–8% of the final fitting value of τ_d or K, the fitting calculation could not reach convergent results due to the presence of measurement errors. Thus, they used 10% of final fitting values of the two parameters as the tolerance in the calculations, which meant that the relative difference of the final fitting value with respect to each of the calculation values (such as $\Delta\tau$ with respect to τ_{d12} or τ_{d23} or τ_{d31}) was below 10%.

9.5 Influence of Shear Stress on Bone Piezovoltage

As a continuation of Section 9.4, this section describes measurements of the piezovoltage of bone between two surfaces of a bone beam under bending deformation. It summarizes the development reported in Fu et al. [23], whose study found that the sign of piezovoltages depended on shear stress only and was not sensitive to normal stress.

9.5.1 Methods

In the experiment, Fu et al. harvested cortical bone materials from the mid-diaphysis of dry bovine (age 2–3 years) tibias and machined them into beam samples with their axes aligned with the longitudinal direction of the diaphysis (Figure 9.14a). Ten samples were prepared with the dimension range

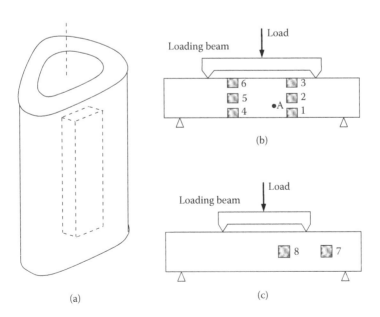

(a)

(b)

(c)

FIGURE 9.14
Sample and electrode arrangements.

TABLE 9.4

Dimension Range of Samples

Span (mm)	Width (mm)	Height (mm)
85	4.0 ± 1.0	19 ± 3.0

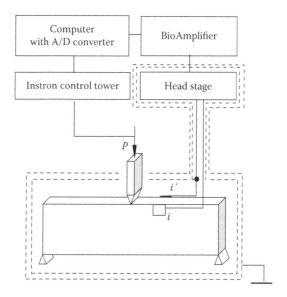

FIGURE 9.15
Setup of the test system for three-point bending test.

shown in Table 9.4. Conductive silver adhesive (5001, SPI, United States) was painted on both sides of the specimens as electrodes with dimensions of 3 mm × 3 mm. Two types of specimen were used with different electrode distributions. Figure 9.14(b) shows type A with six pairs of electrodes and Figure 9.14(c) shows type B with two pairs of electrodes. Two electrodes on both lateral sides of a sample comprised a pair of electrodes located at the same height of both lateral sides, as shown in Figure 9.14.

The electrodes attached to the type A sample are denoted by the numbers 1 to 6 on one side and 1' to 6' for the corresponding electrodes on the other side, facing the medullary cavity. Then, 1–1' to 6–6' represent the six pairs of electrodes, respectively; similarly, 7–7' to 8–8' represent the two pairs of electrodes on the type B sample. When the samples were subjected to four-point bending, all six pairs of electrodes on the type A samples experienced pure bending deformation, whereas on the type B samples only electrodes 8–8' experienced pure bending deformation.

The experimental setup of the measurement system used in Fu et al. [23] is similar to that shown in Figure 9.10, except that a pair of electrodes *i–i'* was used, as shown in Figure 9.15. Taking as reference an electrode on the lateral side facing the medullary cavity, such as 3′, the measured piezo-voltage between the pair of electrodes 3–3′ was input into a bioamplifier (BMA-931, CWE Inc., United States)via an ultrahigh-input impedance (over 1012 Ω) head stage (Super Z, CWE Inc., United States). The amplified volt-age signals were then recorded by a computer in the measurement sys-tem. Mechanical loadings were applied using an Instron 1343 closed-loop servo-hydraulic machine controlled by an 8800 Control Tower, and the loading signals in the control tower were also input into the computer for recording.

The testing sample and the head stage were enclosed in a double electro-magnetic shield box with the outer shield connecting to earth and the inner to the head stage common terminal. This arrangement kept the electric field constant around the sample [4].

A trapezoidal loading profile was used, with a loading (or unloading) rate of 600 N s^{-1}, load holding time of 6 s, and a holding load of 300 N, which pro-duced maximum tensile (or compressive) stresses from 24 to 45 MPa in the samples—well below the tensile strength of 50–100 MPa of cortical bovine bone [36]. The trapezoidal loading waveform is shown in Figure 9.16(a). Figure 9.16(b) shows a typical piezovoltage waveform between electrodes 3–3′ under three-point bending. The waveform has two pulses with different signs, corresponding to the loading and unloading processes, respectively. The peak of the first pulse corresponds to the loading end point and the second pulse peak corresponds to the unloading end point. This shows that once the loading or unloading process ended, the corresponding piezovolt-age began to decrease toward zero. In other words, the polarized charges or piezovoltages appeared on the bone surfaces in pulse form as the bone deformation varied with loading [4].

9.5.2 Results

Figure 9.17 shows a group of typical results of a type A sample. Figure 9.17(a)–(e) corresponds to the arrangement of electrodes on the bone samples. There are two plots of piezovoltage versus time in each figure. The curves I plot represents the result from the three-point bending test and the curves II represents the result from the four-point bending test.

Fu et al. [23] found that all five type A samples showed similar plots of piezovoltage versus time under three-point bending. However, in the plots from the four-point bending, the shapes were similar but the signs were opposite. Tables 9.5 and 9.6 show all the voltage peak values of the sample type A plots, under three- and four-point bending tests.

Figure 9.18 shows the results of piezovoltage versus time for the type B samples under the loading profile shown in Figure 9.16. It can be seen from Figure 9.18 that the signs of the piezovoltage are the same as those shown in Figure 9.17. Table 9.7 lists all the peak values of piezovoltage for the type B samples.

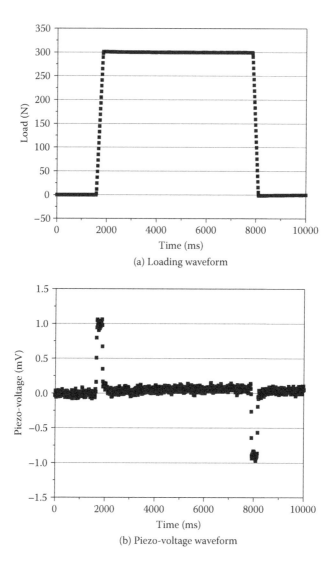

FIGURE 9.16
Loading and piezovoltage waveforms.

9.5.3 Discussion

The results in the preceding section indicate that in the case of three-point bending, the signs of the piezovoltage at electrodes 1–1′ ~ 3–3′ were opposite to those at electrodes 4–4′ ~ 6–6′, and the signs at electrodes 7–7′ ~ 8–8′ of the type B sample were the same as those at electrodes 1–1′ ~ 3–3′ of the type A samples. In the case of four-point bending, the piezovoltages of all the electrodes of the type A samples and electrode 8–8′ of the type B

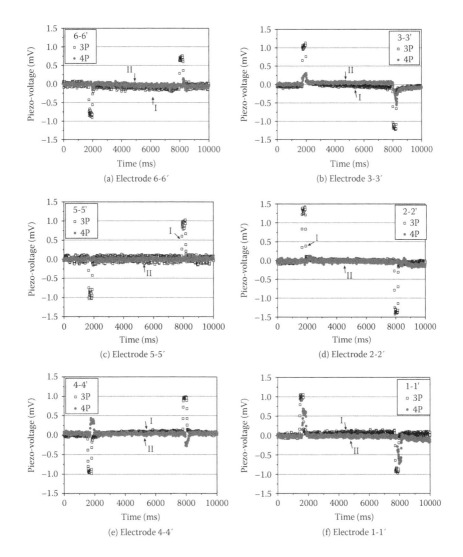

FIGURE 9.17

Typical plots of piezovoltage versus time for type A sample. (I: results for three-point test; II: results for four-point test).

sample were lower than those obtained from the three-point bending test. Moreover, the piezovoltages of the electrodes near the neutral axis in the pure bending zone, such as 2–2′, 5–5′, and 8–8′, were approximately zero. Nevertheless, the piezovoltage signs of electrodes 1–1′, 3–3′, 4–4′, and 6–6′, which were little distant from the neutral axis, were random or irregular. In contrast, the sign and peak of the piezovoltage of electrode 7–7′, which was outside the pure bending zone, were similar to those under the three-point bending test.

TABLE 9.5

Piezovoltage Peak Value (mV) for Type A Sample under Three-Point Bending

Sample Type A	Electrodes					
	1–1′	2–2′	3–3′	4–4′	5–5′	6–6′
1	1.06	1.71	1.23	−0.49	−1.28	−0.94
2	3.27	4.33	1.78	−1.92	−3.94	−2.13
3	1.35	1.32	1.30	−1.35	−1.66	−0.77
4	0.81	2.65	2.47	−1.75	−1.87	−0.79
5	2.39	1.10	1.63	−2.10	−0.99	−1.57

TABLE 9.6

Piezovoltage Peak Value (mV) for Type A Sample under Four-Point Bending

Sample Type A	Electrodes					
	1–1′	2–2′	3–3′	4–4′	5–5′	6–6′
1	0.69	0.10	−0.28	0.43	−0.10	−0.17
2	0.72	0.19	−0.44	0.55	0	−0.31
3	0.40	0	0.56	−0.47	0.14	0
4	0.54	0	0.80	−0.42	0	0.32
5	0.76	−0.15	−0.15	−0.22	−0.20	−0.50

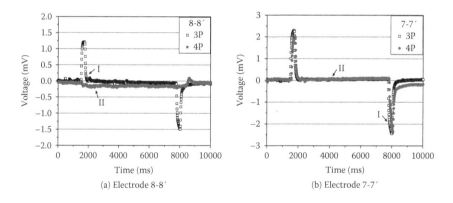

(a) Electrode 8-8′   (b) Electrode 7-7′

FIGURE 9.18

Typical plots of piezovoltage versus time for type B sample. (I: results for three-point test; II: results for four-point test).

Fu et al. [23] summarized their analysis and observations as follows:

1. The differences in the piezovoltage signs between the two groups of electrodes 1–1′ ~ 3–3′ and 4–4′ ~ 6–6′ are due to the fact that normal stress is distributed symmetrically and shear stress is distributed antisymmetrically about the loading axis in the sample under

TABLE 9.7

Piezovoltage Peak Value (mV) for Type B
Sample under Three-and Four-Point Bending

Sample Type B	Electrode 7–7'		Electrode 8–8'	
	3P	4P	3P	4P
1	2.30	1.99	1.20	–0.20
2	1.86	2.03	0.65	0
3	3.07	3.04	1.38	0
4	1.79	1.90	1.76	0.30
5	3.68	3.40	4.01	0.18

three-point bending. This means that the signs of piezovoltage depend on shear stress only. It should be mentioned that the fact that signs of the piezovoltage at electrodes 7–7' and 8–8' were the same as those at electrodes 1–1' ~ 3–3' under three-point bending also supports that conclusion, because electrodes 7–7' and 8–8' are located around the neutral axes and only shear stresses act around them.

2. For both types of samples, there is only normal stress in any cross section in the pure bending zone. Within the pure bending zone, the piezovoltages at all the electrodes become lower, but they do not approach zero except for electrodes 2–2', 5–5', and 8–8', which are located around the neutral axes. It can be concluded that the peak values of the piezovoltages depend mainly on shear stress, although normal stress still contributes to some extent.

3. As can be seen from Tables 9.5–9.7, the peak values at the same electrodes in different samples are relatively irregular. These irregularities occur either between samples or between electrodes. This indicates that the piezoelectricity of bone depends on the hierarchical structure of bone, which might differ in different samples. The important conclusion as to the macroscopic piezoelectric property of bone, however, is that the signs of piezovoltages between two lateral surfaces depend on shear stress only, not on normal stress.

Fu et al. [23] also noted from their analysis that irregularities in the piezovoltage results suggested that the piezoelectric properties of bone are relevant to its microstructure. Cortical bone has been regarded as a hierarchical composite material comprising mineral and organic phases. The mineral phase is mainly composed of crystalline hydroxyapatite and the organic phase consists mainly of collagen, which is the origin of the piezoelectricity [37,38].

Figure 9.19 shows a schematic illustration of collagen fibers [39,40]. A collagen molecule is about 300 nm in length and about 1.5 nm in diameter. There is a 40 nm gap between the ends of collagen molecules in the longitudinal direction, and 27 nm of a 300 nm collagen molecule length overlaps

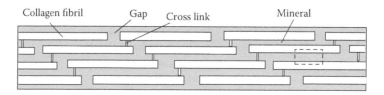

FIGURE 9.19
Schematic illustration of mineral collagen fiber.

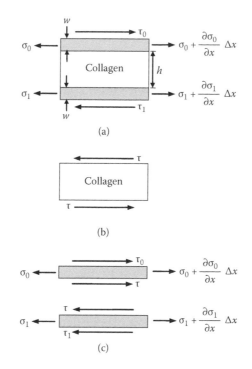

FIGURE 9.20
Schematic presentation of the model.

with the adjacent collagen molecule. There is a covalent cross link in the overlap region connecting adjacent collagen molecules. Collagen molecules are filled and coated by platelet-like tiny mineral crystals, which form the mineralized collagen fibril. A group of collagen fibrils embedded with the mineral crystals forms the hierarchical structure of collagen fiber.

Let the dotted rectangular area in Figure 9.19 represent a segment of mineral collagen. A theoretical model representing the segment is employed in this section [41]. The model has a collagen layer sandwiched between two mineral layers (Figure 9.20a). Although collagen fibrils are oriented in various directions in bone, they are oriented primarily in the direction parallel to the axis of bone diaphysis [42,43] or the axis of the samples. As a force acting on a collagen fiber in any direction can be divided into normal force and shear force, it can be assumed, without loss of generality, that the mineral collagen in the model is parallel to the sample axis. The model can be analyzed using "shear lag" theory [41,44]. Based on this theory, it is assumed that the mineral phase carries normal load only and the collagen phase carries shear load only (Figure 9.20b). The external normal stresses σ_0 and σ_1 can be determined using traditional beam theory. According to the reciprocal theorem of shear stress, the shear stresses τ_0 and τ_1 are equal to the corresponding shear stresses on the cross section. The shear stresses at any point on the cross section can also be determined using traditional beam theory.

Let h and w be the widths of the collagen and mineral components, respectively, and the model thickness be one unit. The force equilibria of the isolated elements (Figure 9.20c) are derived using the analysis method reported in the literature [41]:

$$w\frac{d\sigma_0}{dx} + \tau_0 + \tau = 0 \tag{9.23}$$

$$w\frac{d\sigma_1}{dx} - \tau - \tau_1 = 0 \tag{9.24}$$

Let u_0 and u_1 be displacements of the two mineral components, respectively, and assume the mineral component is linear elastic. Then,

$$\sigma_0 = E\frac{du_0}{dx}, \quad \sigma_1 = E\frac{du_1}{dx}, \quad \tau = G\gamma \tag{9.25}$$

where E is the elastic modulus of the mineral phase, G is the shear elastic modulus, and γ is the shear strain in the collagen layer; γ is also equal to

$$\gamma = \frac{u_1 - u_0}{h} \tag{9.26}$$

Substituting the preceding three equations into Equations (9.23) and (9.24) and then subtracting Equation (9.23) from Equation (9.24) yields

$$wE_m h\frac{d^2\gamma}{dx^2} - 2G\gamma = \tau_0 + \tau_1 \tag{9.27}$$

with the following boundary conditions:

$$\gamma|_{x=\infty} = \text{a bounded value, and also} \quad \gamma|_{x=0} = \text{a bounded value.}$$

Because the origin is a relative position, it can be considered that $\gamma|_{x=0} = 0$. Finally, the solution is

$$\gamma(x) = -\frac{\tau_0 + \tau_1}{2G}\left(1 - e^{-\sqrt{\frac{2G}{wE_m h}}x}\right) \qquad (9.28)$$

The solution (9.28) indicates that the sign of shear strain γ depends entirely on the difference between τ_0 and τ_1 ($\tau_0 > \tau_1$ according to traditional beam theory). The conclusion can be drawn that the shear stress determined by traditional beam theory is responsible for the sign of piezovoltage in bone. Because the exponent term in Equation (9.28) includes elastic modulus E, normal stresses still contribute to the shear strain. However, the normal stress can change only the amplitudes, not the sign of the shear strain. That is, no matter how small τ_0 and τ_1 are, or how large the normal stresses σ_0 and σ_1 are, the signs of shear strain in collagen fibrils depend exclusively on the shear stresses τ_0 and τ_1. This conclusion explains why the signs of piezovoltages remain the same with shear stress under three-point bending.

Fu et al. [23] demonstrated experimentally and theoretically that the signs of piezovoltages of bone under bending deformation depend only on shear stress. It seems doubtful that normal stress contributes only to the amplitude of piezovoltage and not to changing its sign. If normal stress itself can generate a piezovoltage in bone, it must dominate the sign of the voltage, as a voltage is a physical quantity with both magnitude and sign.

A possible reason is the coupling effect between normal stress and shear stress. Macroscopically, only normal stress operates on a cross section of a sample under four-point bending. However, because cortical bone has a hierarchical structure and collagen fibrils are distributed in the mineral matrix in a very complex and random manner, the coupling effect between normal stress and shear stress becomes microscopically stronger. Perhaps the piezovoltages in the pure bending zone of the sample under four-point bending arise from the coupled shear stresses. If so, the contribution of normal stress to piezovoltages in bone still comes substantially from shear stresses.

Other possible contributors to bone piezoelectricity are the cross links, which are covalent bonds between two adjacent collagen molecules (Figure 9.19). Pollack et al. [45] found experimentally that the piezoelectricity of bone increased with an increasing number of cross links. Minary-Jolandan and Yu [38] formed the opinion that cross links enable collagen molecules to transmit mechanical forces to neighboring collagen molecules. In terms of their transmission function, the cross links are affected by shear stress. The contribution of piezosignals from the cross links to the piezovoltages from the bending experiment is an interesting research issue.

References

1. Ho S. P., Goodis H., Balooch M., Nonomura G., Marshall S. J., Marshall G. The effect of sample preparation technique on determination of structure and nano-mechanical properties of human cementum hard tissue. *Biomaterials* 25 (19): 4847–4857 (2004).
2. Mullins L. P., Bruzzi M. S., McHugh P. E. Measurement of the microstructural fracture toughness of cortical bone using indentation fracture. *Journal of Biomechanics* 40 (14): 3285–3288 (2007).
3. Yin L., Venkatesan S., Kalyanasundaram S., Qin Q. H. Effect of microstructure on micromechanical performance of dry cortical bone tissues. *Materials Characterization* 60 (12): 1424–1431 (2009).
4. Hou Z. D., Fu D. H., Qin Q. H. An exponential law for stretching–relaxation properties of bone piezovoltages. *International Journal of Solids and Structures* 48 (3–4): 603–610 (2011).
5. Yin L., Venkatesan S., Kalyanasundaram S., Qin Q. H. Influence of enzymatic maceration on the microstructure and microhardness of compact bone. *Biomedical Materials* 5 (1): 015006 (2010).
6. Mairs S., Swift B., Rutty G. N. Detergent—An alternative approach to traditional bone cleaning methods for forensic practice. *American Journal of Forensic Medicine and Pathology* 25 (4): 276–284 (2004).
7. Belfie D. J., Clark J. M. Enzymatic preparation of allograft bone. *Clinical Research* 40 (1): A134–A134 (1992).
8. Mooney M. P., Kraus E. M., Bardach J., Snodgass J. I. Skull preparation using the enzyme-active detergent technique. *Anatomical Record* 202 (1): 125–129 (1982).
9. Rennick S. L., Fenton T. W., Foran D. R. The effects of skeletal preparation techniques on DNA from human and non-human bone. *Journal of Forensic Sciences* 50 (5): 1016–1019 (2005).
10. Steadman D. W., DiAntonio L. L., Wilson J. J., Sheridan K. E., Tammariello S. P. The effects of chemical and heat maceration techniques on the recovery of nuclear and mitochondrial DNA from bone. *Journal of Forensic Sciences* 51 (1): 11–17 (2006).
11. Athanasiou K. A., Zhu C. F., Lanctot D. R., Agrawal C. M., Wang X. Fundamentals of biomechanics in tissue engineering of bone. *Tissue Engineering* 6 (4): 361–381 (2000).
12. Nalla R. K., Kinney J. H., Ritchie R. O. Mechanistic fracture criteria for the failure of human cortical bone. *Nature Materials* 2 (3): 164–168 (2003).
13. Nalla R. K., Kruzic J. J., Kinney J. H., Ritchie R. O. Mechanistic aspects of fracture and R-curve behavior in human cortical bone. *Biomaterials* 26 (2): 217–231 (2005).
14. Taylor D., Hazenberg J. G., Lee T. C. Living with cracks: Damage and repair in human bone. *Nature Materials* 6 (4): 263–268 (2007).
15. Lawn B. *Fracture of brittle solids*, 2nd ed. Cambridge: Cambridge University Press (1993).
16. Johnson W. M., Rapoff A. J. Microindentation in bone: Hardness variation with five independent variables. *Journal of Materials Science—Materials in Medicine* 18 (4): 591–597 (2007).

17. Evans G. P., Behiri J. C., Currey J. D., Bonfield W. Microhardness and young modulus in cortical bone exhibiting a wide-range of mineral volume fractions, and in a bone analog. *Journal of Materials Science—Materials in Medicine* 1 (1): 38–43 (1990).
18. Riches P. E., Everitt N. M., Heggie A. R., McNally D. S. Microhardness anisotropy of lamellar bone. *Journal of Biomechanics* 30 (10): 1059–1061 (1997).
19. Mahoney E., Holt A., Swain M., Kilpatrick N. The hardness and modulus of elasticity of primary molar teeth: An ultra-micro-indentation study. *Journal of Dentistry* 28 (8): 589–594 (2000).
20. Aschero G., Gizdulich P., Mango F. Statistical characterization of piezoelectric coefficient d(23) in cow bone. *Journal of Biomechanics* 32 (6): 573–577 (1999).
21. Kalinin S. V., Rodriguez B. J., Jesse S., Thundat T., Gruverman A. Electromechanical imaging of biological systems with sub-10 nm resolution. *Applied Physics Letters* 87 (5): 053901 (2005).
22. Minary-Jolandan M., Yu M. F. Nanoscale characterization of isolated individual type I collagen fibrils: Polarization and piezoelectricity. *Nanotechnology* 20 (8): 085706 (2009).
23. Fu D. H., Hou Z. D., Qin Q. H., Xu L., Zeng Y. J. Influence of shear stress on behaviors of piezoelectric voltages in bone. *Journal of Applied Biomechanics* (in press) (2012).
24. An Y. H., Draughn R. A. *Mechanical testing of bone and the bone–implant interface.* Boca Raton, FL: CRC Press (2000).
25. Xu H. H. K., Smith D. T., Jahanmir S., Romberg E., Kelly J. R., Thompson V. P., Rekow E. D. Indentation damage and mechanical properties of human enamel and dentin. *Journal of Dental Research* 77 (3): 472–480 (1998).
26. Everitt N. M., Rajah S., McNally. D. S. Bone recovery following indentation. *Journal of Bone Joint Surgery*, British vol. 88-B: 398 (2006).
27. Fan Z. F, Rho J. Y. Effects of viscoelasticity and time-dependent plasticity on nanoindentation measurements of human cortical bone. *Journal of Biomedical Materials Research Part A* 67A (1): 208–214 (2003).
28. Ramrakhiani M., Pal D., Murty T. S. Micro-indentation hardness studies on human bones. *Acta Anatomica* 103 (3): 358–362 (1979).
29. Singh S., Saha S. Electrical properties of bone—A review. *Clinical Orthopaedics and Related Research* 186: 249–271 (1984).
30. Johnson M. W., Chakkalakal D. A., Harper R. A., Katz J. L. Comparison of the electro-mechanical effects in wet and dry bone. *Journal of Biomechanics* 13 (5): 437–442 (1980).
31. Fu D. H., Hou Z. D., Qin Q. H., Jiang X. G. Analysis of the waveforms of piezoelectric voltage of bone. *Journal of Tianjin University, Science and Technology* 39 (SUPPL.): 349–353 (2006).
32. Yamashita J., Li X. O., Furman B. R., Rawls H. R., Wang X. D., Agrawal C. M. Collagen and bone viscoelasticity: A dynamic mechanical analysis. *Journal of Biomedical Materials Research* 63 (1): 31–36 (2002).
33. Fondrk M., Bahniuk E., Davy D. T., Michaels C. Some visco-plastic characteristics of bovine and human cortical bone. *Journal of Biomechanics* 21 (8): 623–630 (1988).
34. Hoc T., Henry L., Verdier M., Aubry D., Sedel L., Meunier A. Effect of microstructure on the mechanical properties of Haversian cortical bone. *Bone* 38 (4): 466–474 (2006).

35. Williams G., Watts D. C. Non-symmetrical dielectric relaxation behaviour arising from a simple empirical decay function. *Transactions of the Faraday Society* 66 (565P): 80–85 (1970).
36. Cowin S. C. The mechanical and stress adaptive properties of bone. *Annals of Biomedical Engineering* 11 (3–4): 263–295 (1983).
37. Steinber Me., Bosch A., Schwan A., Glazer R. Electrical potentials in stressed bone. *Clinical Orthopaedics and Related Research* 61: 294–300 (1968).
38. Minary-Jolandan M., Yu M. F. Uncovering nanoscale electromechanical heterogeneity in the subfibrillar structure of collagen fibrils responsible for the piezoelectricity of bone. *ACS Nano* 3 (7): 1859–1863 (2009).
39. Fratzl P., Weinkamer R. Nature's hierarchical materials. *Progress in Materials Science* 52 (8): 1263–1334 (2007).
40. Jager I., Fratzl P. Mineralized collagen fibrils: A mechanical model with a staggered arrangement of mineral particles. *Biophysical Journal* 79 (4): 1737–1746 (2000).
41. Wang X. D., Qian C. J. Prediction of microdamage formation using a mineral–collagen composite model of bone. *Journal of Biomechanics* 39 (4): 595–602 (2006).
42. Ascenzi M. G., Lomovtsev A. Collagen orientation patterns in human secondary osteons, quantified in the radial direction by confocal microscopy. *Journal of Structural Biology* 153 (1): 14–30 (2006).
43. Bills P. M., Lewis D., Wheeler E. J. Mineral–collagen orientation relationships in bone. *Journal of Crystallographic and Spectroscopic Research* 12 (1): 51–53 (1982).
44. Kotha S. P., Guzelsu N. The effects of interphase and bonding on the elastic modulus of bone: Changes with age-related osteoporosis. *Medical Engineering & Physics* 22 (8): 575–585 (2000).
45. Pollack S. R., Korostoff E., Sternberg M. E., Koh J. Stress-Generated potentials in bone: Effects of collagen modifications. *Journal of Biomedical Materials Research* 11 (5): 677–700 (1977).

Appendix A: Bone Types Based on Pattern of Development and Region*

In Section 1.1, bone types based on a macroscopic approach, microscopic observation, and geometric shape are described. Those typologies are used as the major reference throughout the book. For completeness, bone types based on pattern of development and region are presented here.

A.1 Bone Types Based on Pattern of Development

On the basis of the pattern of development of a bone, bones can be classified into three categories. In fact, bone formation (osteogenesis) begins during prenatal development and persists through adulthood. Bones of infants and children are softer than those of adults because they have not yet been ossified (the process of synthesizing cartilage into bone). Osteogenesis occurs in two ways: by intramembranous ossification and by endochondral ossification. Both types form bone by replacing existing cartilage, but they differ in the way they go about it. Two types of cells that are of great importance in the process are osteoblasts and osteoclasts. Osteoblasts, used mainly in intramembranous ossification, are the specialized cells in bone tissue that deposit calcium into the protein matrix of bone (collagen). Osteoclasts, used in endochondral ossification, dissolve calcium previously stored away in bone and carry it to tissues whenever needed.

A.1.1 Membranous Bones

Bones formed during intramembranous ossification are called membranous bones or, occasionally, dermal bone. These bones ossify from mesenchymal condensation in intrauterine life.

It can be observed under a microscope that membranous bones first appear as flat, membrane-like layers of early connective tissue. These layers are provided with a constant flow of nutrient blood supply by networks of blood vessels formed between the layers. Early connective tissue cells first arrange themselves among the layers and then differentiate into bone-forming cells called osteoblasts. The osteoblasts then remove calcium from the blood and

---

* http://en.wikipedia.org/wiki/Bone (accessed March 2011).

deposit it in the bone matrix (the cartilage). As a result, layers of spongy bone are formed around the original cartilage.

Later in development, spaces among the spongy bone are filled with bone matrix and become compact bone. Osteoblasts continue to deposit calcium supplements into the matrix until it is totally surrounded by bone. Once this occurs, the osteoblasts are considered to be encased in a lacuna and are now called osteocytes. The original connective cells first formed around the network of blood vessels are now called the periosteum. Osteoblasts still not isolated in lacunae can emerge from beneath the layer of compact bone and form layers of spongy bone over compact bone. Examples of membranous bones are bones of the skull and facial bones.

A.1.2 Cartilaginous Bones

Cartilaginous bones ossify in cartilage (intracartilaginous or endochondral ossification) and are thus derived from preformed cartilaginous models. Endochondral ossification forms bone by replacing a cartilaginous model, or precursor, that appeared there earlier in embryonic development. The cartilaginous models first undergo rapid changes as the connective tissue cells enlarge, which in turn destroys the surrounding matrix. Soon after, the connective tissue cells die.

While the cells disintegrate, a periosteum is formed on the outside of the developing structure (a membrane with many blood vessels). Next, blood vessels and undifferentiated cells raid into the disintegrating tissue. Certain connective tissue cells differentiate and form spongy bone around the previous template of cartilage. Examples of cartilaginous bone are the bones of limbs, the vertebral column, and the thoracic cage. A defect in endochondral ossification causes a common type of dwarfism called achondroplasia, in which the limbs are short but the trunk is normal. It is transmitted as a Mendelian dominant character.

A.1.3 Membrocartilaginous Bones

There is a third type of ossification, in which a bone ossifies partly from membrane and partly from cartilage. These bones are known as membrocartilaginous bones. In the process of bone formation, thickness in cartilage bones is achieved by intramembranous ossification. Just beneath the layer of periosteum, yet above the newly developed spongy bone, compact bone is formed and hardened with the help of osteoblasts filling portions of the porous spongy bone with calcium phosphate crystals (apatite). Sometimes compact bone is formed on the surfaces of existing bone tissue and must be eroded away by specialized cells called osteoblasts. The crystals of apatite extracted from the bone tissue are delivered to blood and tissues on demand. Examples of this class of bone include the clavicle, mandible, occipital, temporal, and sphenoid bones.

A.2 Bone Types Based on Region

Based on the region where bones locate, we have two bone types. The human skeleton can be divided into two main parts; the first part is the axial skeleton and second is the appendicular skeleton.

The axial skeleton consists of the bones in the head and trunk of the human body. This area can be further divided into four main parts: the skull, the hyoid bone of the throat, the chest, and the vertebral column. The appendicular skeleton consists of 126 bones in the human body that make motion possible and protect the organs of digestion, excretion, and reproduction. The word "appendicular" refers to an appendage or anything attached to a major part of the body, such as the upper and lower extremities. The axial skeleton and the appendicular skeleton together form the complete skeleton.

A.2.1 Bones of the Axial Skeleton

The axial skeleton forms the central axis of the body. It consists of the 80 bones along the central axis of the human body. It is composed of four parts: the human skull, the hyoid bone of the throat, the vertebral column, and the thoracic cage

A.2.1.1 Skull

The skull consists of 28 different bones (including the ossicles of the ear). The bones of the skull can be divided into three main groups: the cranium, the facial bones, and the middle ear.

- Cranium (8)
  - Frontal bone (1) forms the forehead and the superior surface of the orbits.
  - Parietal bones (2) are found on both sides of the skull, posterior to the frontal bone. They are bones in the human skull that, when joined together, form the sides and roof of the cranium. Each bone is roughly quadrilateral in form and has two surfaces, four borders, and four angles.
  - Occipital bone (1) forms the posterior and inferior portions of the cranium.
  - Temporal bones (2) are found below the parietal bones, contributing to the sides and base of the cranium.
  - The sphenoid bone (1) forms part of the floor of the cranium. The sphenoid bone is an unpaired bone situated at the base of the skull in front of the temporal bone and basilar part of the occipital bone. It is one of the seven bones that articulate to form

the orbit. Its shape somewhat resembles that of a butterfly or bat with its wings extended.

- The ethmoid bone (1) is a bone in the skull that separates the nasal cavity from the brain. It is located at the roof of the nose, between the two orbits. The cubical bone is light in weight due to its spongy construction. The ethmoid bone is one of the bones that make up the orbit of the eye. The ethmoid has three parts: the cribriform plate, the ethmoidial labyrinth, and the perpendicular plate.

- Facial (14)

  - Maxillary bones (2) are the bones of the upper jaw that serve as a foundation of the face and support the orbits. They form the floor and medial portion of the orbit rim, walls of the nasal cavity, and the anterior roof of the mouth (hard palate).

  - Zygomatic bones (2) are found on each side of the skull, articulating with the frontal bone and the maxilla to complete the lateral wall of the orbit. Along the lateral margin, each gives rise to a slender bony extension that curves laterally and posteriorly to meet a process from the temporal bone, together forming the zygomatic arch.

  - Palatine bones (2) form the posterior surface of the hard palate. The superior surfaces of each horizontal portion contribute to the floor of the nasal cavity. The superior tip of the vertical portion of each forms a small portion of the inferior wall of the orbit.

  - The mandible (1) forms the lower jaw. It is a bone forming the skull with the cranium.

  - Lacrimal bones (2) are located within the orbit on its medial surface and articulate with the frontal, ethmoid, and maxillary bones.

  - Nasal bones (2) are two small, oblong bones, varying in size and form in different individuals; they are located side by side in the middle and upper part of the face. The nasal bones form the bridge of the nose and articulate with the superior frontal bone and the maxillary bones.

  - Inferior nasal conchae (2) are among the turbinates in the nose. They extend horizontally along the lateral wall of the nasal cavity and consist of a lamina of spongy bone, curled upon itself like a scroll. Each inferior nasal concha is considered a facial pair of bones since they arise from the maxillary bones and project horizontally into the nasal cavity.

  - The vomer (1) is one of the unpaired facial bones of the skull. It is located in the midsagittal line and articulates with the sphenoid, the ethmoid, the left and right palatine bones, and the left and right maxillary bones. It supports a prominent partition that forms part of the nasal septum.

- Middle ear (6)
  - The middle ear contains the auditory ossicles—tiny bones in the middle ear of vertebrates that connect the eardrum to the inner ear. They transmit vibrations of the eardrum caused by sound waves in the air to the fluid of the inner ear via the oval window. Three ossicles—the malleus (hammer), incus (anvil), and stapes (stirrup)—are present in the mammalian ear. They form a lever system, diminishing the force of sound waves and increasing the force on the inner ear. A single ossicle occurs in the ear of birds, amphibians, and reptiles.
  - The malleus (2) is the largest ossicle and has a head, a neck, a long process or handle, an anterior process, and a lateral process. The head is rounded and lies within the epitympanic recess. It articulates posteriorly with the incus. The neck is the restricted part below the head. The handle passes downward and backward and is firmly attached to the medial surface of the tympanic membrane (eardrum). It can be seen through the tympanic membrane on otoscopic examination. The anterior process is a spicule of bone that is connected to the anterior wall of the tympanic cavity by a ligament. The lateral process projects laterally and is attached to the anterior and posterior malleolar folds of the tympanic membrane.
  - The incus (2) is the anvil-shaped small bone or ossicle in the middle ear. It attaches the malleus to the inner bone (stapes). The incus transmits sound vibrations from the malleus to the stapes.
  - The stape (2) is the stirrup-shaped small bone or ossicle in the middle ear that is attached through the incudostapedial joint to the incus laterally and to the fenestra ovalis, the "oval window, " medially.

A.2.1.2 Hyoid Bone (1)

The hyoid bone is a U-shaped bone that hangs below the skull, suspended by ligaments from the styloid processes of the temporal bones, and serves as a base for muscles associated with the tongue and larynx. At rest, it lies at the level of the base of the mandible in the front and the third cervical vertebra behind. Unlike other bones, the hyoid is only distantly articulated to other bones by muscles or ligaments.

A.2.1.3 Vertebral Column (26)

The vertebral column forms the central part of the skeleton. It supports the skull and protects the spinal cord. It also serves as attachment for the ribs and the pectoral and pelvic girdles. The vertebral column consists of separate bones, the vertebrae. The vertebrae are arranged above each other. Because the separate vertebrae are attached to each other by means of fibrous cartilaginous discs, they form a flexible column. Each vertebra has articular surfaces above and below, which allow articulation movement between them. The vertebral column of 26 vertebrae is divided into five regions according to their position and structure. There are seven cervical (neck) vertebrae, twelve thoracic (chest) vertebrae, five lumbar vertebrae, one fused sacral vertebra, and one coccyx.

- Cervical vertebrae. The neck region consists of seven cervical vertebrae. These are the smallest vertebrae in the vertebral column. The first two cervical vertebrae are known as the atlas and axis. They are specially adapted to support the skull and to enable it to move. They differ from the structure of the typical vertebra in certain respects.
- Thoracic vertebrae. There are 12 thoracic vertebrae. The centrum is large and sturdy and the neural spines are long and directed downward. The long neural spines form an anchorage for the muscles and ligaments that support the head and neck. The head (or capitulum) of each of the first 10 pairs of ribs fits into and articulates with the semicircular facet, which is situated between two successive centra (i.e., between the inferior surface of one and the superior surface of the next centrum). These facets occur on both sides of the centrum. The tubercle of the rib articulates with the facet at the tip of the transverse process.
- Lumbar vertebrae. These 5 vertebrae are the largest and strongest in the vertebral column. The transverse processes are very long for the attachment of the powerful back muscle that maintains the posture and flexes the spine in movement.
- The sacrum (1) forms the posterior wall of the pelvis. The sacrum is a large, triangular bone at the base of the spine and at the upper and back part of the pelvic cavity, where it is inserted like a wedge between the two hip bones. Its upper part connects with the last lumbar vertebra and the bottom part with the coccyx (tailbone).
- The coccyx (1) is one mass of four to five small coccygeal vertebrae that have fused into one, commonly called the tailbone.

A.2.1.4 Thoracic Cage (25)

The human rib cage, also known as the thoracic cage, is a bony and cartilaginous structure that surrounds the thoracic (chest) cavity and supports the pectoral (shoulder) girdle, forming a core portion of the human skeleton. A typical human ribcage consists of 14 true ribs, 10 false ribs, and one sternum.

- True ribs (14) consist of seven pairs of bone that reach directly to the sternum with hyaline cartilage called costal cartilage. They are connected to the sternum by separate cartilaginous extensions (costal cartilages).
- False ribs (10) are the five sets of ribs below the top seven pairs of true ribs. A rib is considered to be "false" if it has no direct attachment to the sternum, also known as the breast bone. Among them, three pairs of false ribs (vertebrochondral ribs) do not attach to the sternum. Rather, they connect (with costal cartilage) to the rib directly above them. Two pairs of false ribs (floating ribs or vertebral ribs) do not attach to anything at their anterior ends.
- The sternum (1) is a long, flat, bony plate shaped like a capital "T" located anteriorly to the heart in the center of the thorax (chest). It connects the rib bones via cartilage, forming the anterior section of the rib cage with them, and thus helps to protect the lungs, heart, and major blood vessels from physical trauma. Although it is fused, the sternum can be subdivided into three regions: the manubrium, the body, and the xiphoid process.

A.2.2 Bones of the Appendicular Skeleton

The appendicular skeleton (126 bones) consists of the pectoral girdles (4); the upper limbs (60); the pelvic girdle (2), which is attached to the pectoral (shoulder) girdle; and the lower limbs (60) attached to the pelvic girdle. Their functions are to make locomotion possible and to protect the major organs of locomotion, digestion, excretion, and reproduction. Like the bones of the axial skeleton, the bones of the limbs and their girdles are preformed in cartilage. Cartilage is a dense connective tissue that has many properties similar to bone. The strength of both relies on the secretion of a rigid extracellular matrix.

Though cartilage precedes bone in evolution as well as ontology, living creatures with completely cartilaginous skeletons are known as chondrichthyans and have this as a derived condition, having lost the ability to make bone. There are no land-dwelling animals with a cartilaginous skeleton. There are many reasons for this: While rigid, cartilage does not resist tension as well as bone does, is more flexible and elastic (which is a problem when dealing with gravity), and is incapable of remodeling. That said, it is perhaps this flexibility (as well as its lighter weight) that makes cartilage well suited to aquatic environments where the animal is not supporting itself against gravity. This includes the environment of the womb in amniotes and the water environment in which anamniotes lay their eggs.

- The pectoral girdle (4) consists of two shoulder blades (scapulae) and two collar bones (clavicles). These bones articulate with one another, allowing some degree of movement. Each shoulder blade sits at rest over the ribs of the back, and the collarbone attaches in the front of the body with the sternum.

- The scapula (2) is commonly called the shoulder blade. It is supported and positioned by the skeletal muscles. The scapula has no bony or ligamentous bonds to the thoracic cage, but it is extremely important for muscle attachment.

- The clavicle (2) is commonly known as the collarbone. It articulates with the manubrium of the sternum and is the only direct connection between the pectoral girdle and the axial skeleton.

- The upper limb or upper extremity (60) is the region in an animal extending from the deltoid region to the hand, including the arm, axilla, and shoulder. It may be divided into five main regions: an upper arm bone (humerus), the forearm (radius and ulna), the wrist (carpals), the metacarpals, and the phalanges.

  - The humerus (2) is the bone of the shoulder and arm. It articulates with the scapula at the shoulder and with the radius and ulna at the elbow. The distal end of the humerus includes the lateral and medial epicondyles and a condyle consisting of the capitulum and trochlea. The lateral epicondyle gives origin to the supinator and to the extensor muscles of the forearm. The capitulum articulates with the head of the radius. The trochlea is a pulley-shaped projection that articulates with the trochlear notch of the ulna. It is set obliquely, so that a "carrying angle" exists between the arm and the extended and supinated forearm. Radial and coronoid fossae are situated anterior and superior to the capitulum and trochlea, respectively. A deeper olecranon fossa is located posteriorly, superior to the trochlea. The medial epicondyle gives origin to the flexor muscles of the forearm. The ulnar nerve lies in a groove posterior to the medial epicondyle and is palpable there (the "funny bone"). The medial epicondyle gives an indication of the direction in which the head of the humerus is pointing in any given position of the arm. The distal end of the humerus is angled anteriorward, and a decrease in the normal angulation suggests a supracondylar fracture.

  - The radius (2) lies along the lateral side (or thumb side) of the forearm. The proximal end articulates with the humerus, the medial aspect with the ulna, and the distal end with the carpus. The distal end of the radius terminates in the styloid process laterally. The process is palpable between the extensor tendons of the thumb. It gives attachment to the radial collateral ligament. The styloid process of the radius is about 1 cm distal to that of the ulna. This relationship is important in the diagnosis of fractures and in the verification of their correct reduction. On its medial side, the distal end of the radius has an ulnar notch, for articulation with the head of the ulna. At about the middle of the convex dorsal aspect of the distal end of the radius, a small prominence, the dorsal tubercle, may be felt. The inferior surface of the distal end articulates with the lunate (medial) and the scaphoid (lateral). A fall on the outstretched hand may result in a fracture of the distal end of the radius, in which the distal fragment is displaced posteriorly and generally becomes impacted, bringing the styloid processes of the radius and ulna to approximately the same horizontal level.

- The ulna (2) forms the medial support of the forearm. It is longer than and medial to the radius. It articulates with the humerus proximally, the radius laterally, and the articular disc distally. The proximal end includes the olecranon and the coronoid process. The olecranon is the prominence of the posterior elbow that rests on a table when a person leans on the elbow. The lateral epicondyle, the tip of the olecranon, and the medial epicondyle are in a straight line when the forearm is extended, but form an equilateral triangle when the forearm is flexed. The superior aspect of the olecranon receives the insertion of the triceps. The posterior aspect, covered by a bursa, is subcutaneous. The anterior part of the olecranon forms a part of the trochlear notch, which articulates with the trochlea of the humerus. The coronoid process, which completes the trochlear notch, projects anteriorward and engages the coronoid fossa of the humerus during flexion. It is prolonged inferiorward as a rough area termed the tuberosity of the ulna. The radial notch is on the lateral aspect of the coronoid process and articulates with the head of the radius.

- The carpal bones (16) are eight pairs of bones of the wrist and consist of (a) four proximal bones (scaphoid, lunate, triangular or triquetral, and pisiform), and (b) four distal bones (trapezium, trapezoid, capitate, and hamate). The posterior aspect of the intact carpus is convex and the anterior aspect is concave, where it is bridged by the flexor retinaculum to form the carpal canal or tunnel for the flexor tendons and the median nerve. Hence, the posterior surfaces of the carpals are generally larger than the anterior, with the exception of the lunate, where the converse holds. The flexor retinaculum extends between the scaphoid and trapezium laterally and the triquetrum and hamate medially. These four bones can be distinguished by deep palpation. The scaphoid has a tubercle on its anterior side that can be felt under cover of and lateral to the tendon of the flexor carpi radialis. A fall on the outstretched hand may result in fracture of the scaphoid, generally across its "waist." In some fractures the blood supply of the proximal fragment may be compromised, resulting in aseptic necrosis. The lunate is broader on the anterior than the posterior side. Anterior dislocation of the lunate is a fairly common injury of the wrist.

- The metacarpals (10) consist of five pairs of bones that articulate with the distal carpal bones forming the palm of the hand. The carpus is connected to the phalanges by five metacarpal bones, referred to collectively as the metacarpus. They are numbered from one to five, from the thumb to the little finger. The first is the shortest and the second the longest. They contribute to the palm, and their posterior aspects can be felt under cover of the extensor tendons. Each metacarpal is technically a long bone, consisting of a base proximally, a shaft, and a head distally. The base articulates with the carpus and, except for that of the first, with the adjacent metacarpal(s) also. The base of the first metacarpal has a saddle-shaped facet for the trapezium. The head of each metacarpal articulates with a proximal phalanx and forms a knuckle of the fist.

- The phalanges (28) consist of 14 pairs of finger bones. Four fingers contain three phalanges while the pollex (thumb) has only two. They are designated proximal, middle, and distal. Each phalanx is technically a long bone, consisting of a base proximally, a shaft, and a head distally. The base of a proximal phalanx articulates with the head of a metacarpal, and the head of the phalanx presents two condyles for the base of a middle phalanx. Similarly, the head of a middle phalanx presents two condyles for the base of a distal phalanx. Each distal phalanx ends in a rough expansion termed its tuberosity. Sesamoid bones are found related to the anterior aspects of some of the metacarpophalangeal and interphalangeal joints. Two located anterior to the head of the first metacarpal are almost constant.

- The pelvic girdle (2) articulates with the femur. The bony pelvis is formed by the hip bones in front and at the sides and by the sacrum and coccyx behind. The hip bone (also, innominate bone or coxal bone) is a large, flattened, irregularly shaped bone, constricted in the center and expanded above and below. It has one of the few ball-and-socket synovial joints in the body: the hip joint.

- Lower limbs. The skeleton of the lower limb may be divided into five main regions: the upper leg (thigh), the lower leg, the ankle, the arch of the foot, and the toes. The upper leg has a single long bone, the femur, which is the longest bone in the body. The two bones of the lower leg are the tibia (shinbone) in front and the fibula behind. The tibia is the larger of the two and extends from the knee to the ankle. The upper end of the tibia has two articulating facets into which the condyles of the femur fit to form the knee joint. The lower end of the tibia articulates with one of the tarsals to form the ankle joint. The fibula is smaller than the tibia and is situated on the outside and slightly behind it. The upper end articulates with the tibia but does not form part of the knee joint. The lower end forms part of the ankle joint. Regarding the ankle, there are seven short, thick tarsal bones, the largest of which is the heel bone (calcaneum), which presses firmly onto the ground when one stands, walks, or runs. The arch is formed partly by some of the tarsals but mainly by the five long metatarsals, which extend from the tarsals to the toes. The arch of the foot is modified for receiving the weight of the body. There are 14 short phalanges in the toes of each foot. The big toe has two phalanges and the other toes have three each. There are 60 bones in the lower limb:

- The femur (2) is commonly called the thigh bone. It is the longest, strongest, and heaviest bone in the body. Distally, it articulates with the tibia at the knee joint. The head (epiphysis) articulates with the pelvis at the acetabulum.

- The tibia (2), also known as the shinbone or shankbone, is the larger and stronger of the two bones in the leg below the knee in vertebrates and connects the knee with the ankle bones. It is commonly recognized as the strongest weight-bearing bone in the body.

- The fibula (2) is the slender lateral bone of the leg. It does not bear weight. The term "peroneal" is synonymous with fibular and has been used in the past interchangeably. The fibula articulates with the tibia superiorly and with the talus inferiorly and is anchored in between to the tibia by the interosseous membrane. The superior and inferior ends of the bone are palpable, but muscles cover its middle portion. The inferior end of the fibula, or the lateral malleolus, is more prominent and more posterior and extends about 1 cm more distally than the medial malleolus. It articulates with the tibia and with the lateral surface of the talus; the talus fits between the two malleoli. Posteromedially, a malleolar fossa gives attachment to ligaments. Posteriorly, a groove on the lateral malleolus is occupied by the peroneal tendons. The classic (Pott's) fracture at the ankle involves the lower end of the fibula.

- The patella (2) is the kneecap. It is a triangular sesamoid bone embedded in the tendon of insertion of the quadriceps femoris muscle. The superior border of the patella is the base of the triangle, and lateral and medial borders descend to converge at the apex. The patella can be moved from side to side when the quadriceps is relaxed. A part of the quadriceps tendon covers the anterior surface of the bone and is continued, as the patellar ligament, to the tuberosity of the tibia. The patella articulates on its posterior side with the patellar surface of the condyles of the femur. The articular surface of the patella comprises a larger, lateral facet and a smaller, medial one. Lateral dislocation of the patella is resisted by the shape of the lateral condyle of the femur and by the medial pull of the vastus medialis. Excision of the patella results in minimal functional deficiency. The patella ossifies from several centers, which appear during childhood.

- The tarsal bones (14) are seven pairs of bones (talus, calcaneus, navicular, cuboid, and the first, second, and third cuneiform bones). Only the talus articulates with the tibia and fibula. The tarsus is convex superiorly and concave inferiorly. The tarsal bones are the talus, navicular, and three cuneiforms on the medial side, and the calcaneus and cuboid, which are more laterally placed. Accessory ossicles may be found—for example, a fibular sesamoid, the os tibiale externum (near the tuberosity of the navicular and sometimes called the "accessory navicular"), and the os trigonum.

- The metatarsal bones (10) support the sole of the foot and are numbered I to V from medial to lateral, with the distal ends forming the ball of the foot. They are connected to the phalanges by five metatarsal bones, referred to collectively as the metatarsus. Each metatarsal is technically a long bone, consisting of a base proximally, a shaft, and a head distally. Each bone has characteristic features; for example, the first (which carries more weight) is short and thick. The base of the fifth presents a tuberosity, which projects posterolaterally and is palpable at the lateral aspect of the foot.

- The phalanges (28) have the same arrangement as with the fingers and thumb, but with the toes and great toe (hallux). They are designated proximal, middle, and distal. Each phalanx is technically a long bone, consisting of a base proximally, a shaft, and a head distally. Although the phalanges of the foot are shaped differently from those of the hand, their basic arrangement is similar, in that each distal phalanx ends distally in a tuberosity. The middle and distal phalanges of the little toe are often fused. The phalanges usually begin to ossify during fetal life, and centers appear postnatally in the bases of most of them.

Index